WOMEN AND THE ISRAELI OCCUPATION

The state of Israel and the Palestinian nation are at a monumental juncture in their histories. Respective representatives have recognized each other's right to exist, learning to conceive of a new 'other'. Both have a chance to claim a new future, but more than a quarter of a century of occupation has left a permanent mark on Israeli Palestinians, Israeli Jews and Palestinians of the Occupied Territories.

Israel's occupation of the West Bank and the Gaza Strip created a relationship which, similar to that between colonizer and colonized, placed Israeli Jews in the position of the powerful and Palestinians of the Occupied Territories in the position of the powerless.

This dichotomy of more than twenty-six years of occupation has significant social, political, economic, cultural, psychological and moral ramifications for both men and women, both Israelis and Palestinians.

Women and the Israeli Occupation analyses the impact of the unequal occupier/occupied relationship on the lives of Palestinian and Jewish women. Exposing a set of previously unarticulated internal conflicts and differences, the occupation has also reinforced existing loyalties as the different groups of women have moved into public political action, working together to end the occupation. Combining the view of Jews and Palestinians from Israel and Palestinians from the Occupied Territories, this book is the first to discuss the impact of occupation on women. It will prove invaluable to students of Women's Studies, Political Science, Geography and Middle Eastern Studies.

Tamar Mayer is Associate Professor and Chair of Geography at Middlebury College, Vermont.

INTERNATIONAL STUDIES OF WOMEN AND PLACE

Edited by Janet Momsen, *University of California at Davis*
and Janice Monk, *University of Arizona*

The Routledge series of *International Studies of Women and Place* describes the diversity and complexity of women's experience around the world, working across different geographies to explore the processes which underlie the construction of gender and the life-worlds of women.

Other titles in this series:

FULL CIRCLES
Geographies of women over the life course
Edited by Cindi Katz and Janice Monk

'VIVA'
Women and popular protest in Latin America
Edited by Sarah A. Radcliffe and Sallie Westwood

DIFFERENT PLACES, DIFFERENT VOICES
Gender and development in Africa, Asia and Latin America
Edited by Janet H. Momsen & Vivian Kinnard

SERVICING THE MIDDLE CLASSES
Class, gender and waged domestic labour in contemporary Britain
Nicky Gregson and Michelle Lowe

Forthcoming titles:

WOMEN'S VOICES FROM THE RAINFOREST
Janet Gabriel Townsend

WOMEN AND THE ISRAELI OCCUPATION

The politics of change

Edited by Tamar Mayer

London and New York

First published 1994
by Routledge
11 New Fetter Lane, London EC4P 4EE
Simultaneously published in the USA and Canada
by Routledge
29 West 35th Street, New York, NY 10001

Collection as a whole © 1994 Tamar Mayer
Individual chapters © 1994 respective contributors

Typeset in Garamond by
Florencetype Ltd, Stoodleigh, Devon

Printed and bound in Great Britain by
Biddles Ltd, Guildford and King's Lynn

All rights reserved. No part of this book may be reprinted
or reproduced or utilized in any form or by any electronic,
mechanical, or other means, now known or hereafter
invented, including photocopying and recording, or in any
information storage or retrieval system, without permission
in writing from the publishers.

British Library Cataloguing in Publication Data
A catalogue record for this book is available
from the British Library

Library of Congress Cataloging in Publication Data
Women and the Israeli occupation : the politics of change /
edited by Tamar Mayer.
p. cm. – (International studies of women and place)
Includes bibliographical references and index.
1. Women, Palestinian Arab–Social conditions. 2. Women–
Israel–Social conditions. 3. Women, Palestinian Arab–
Political activity. 4. Women–Israel–Political activity.
5. Palestine–Social conditions. 6. Israel–Social conditions.
I. Mayer, Tamar. II. Series.
HQ1728.5.W66 1994
305.42'095694–dc20   94–3886

ISBN 0–415–09545–X   0–415–09546–8 (pbk)

For Joey

CONTENTS

List of tables		ix
List of contributors		x
Acknowledgments		xii
1	Women and the Israeli Occupation: The context Tamar Mayer	1
2	What has the occupation done to Palestinian and Israeli women? A dialogue between *Naomi Chazan* and *Mariam Mar'i*	16
3	Between national and social liberation: The Palestinian women's movement in the Israeli occupied West Bank and Gaza Strip Souad Dajani	33
4	Heightened Palestinian nationalism: Military occupation, repression, difference and gender Tamar Mayer	62
5	Israeli women against the Occupation: Political growth and the persistence of ideology Yvonne Deutsch	88
6	Palestinian women in Israel: Identity in light of the Occupation Nabila Espanioly	106
7	Homefront as battlefield: Gender, military occupation and violence against women Simona Sharoni	121
8	Trends in labor market participation and gender-linked occupational differentiation Moshe Semyonov	138
9	Women street peddlers: The phenomenon of *Bastat* in the Palestinian informal economy Suha Hindiyeh-Mani, Afaf Ghazawneh, and Subhiyyeh Idris	147

CONTENTS

10 Environmental problems affecting Palestinian women under occupation 164
   Karen Assaf

11 A feminist politics of health care: The case of Palestinian women under Israeli occupation, 1979–1982 179
   Elise G. Young

Index 199

TABLES

8.1 Labor force participation of the Arab population of the
West Bank and Gaza Strip 1970–1989   141
8.2 The occupational distribution of the labor force population
in the West Bank and Gaza Strip 1970–1989   143
8.3 Relative odds for men and women employed in selected
occupational categories 1970–1989   144

CONTRIBUTORS

Karen Assaf (Ph.D University of Texas) was one of the first scientists to work on Palestinian water rights in the West Bank and the Gaza Strip. She taught at Birzeit University and helped establish the Arab College of Medical Professions. As the co-director of ASIR, the Arab Scientific Institute for Research and Transfer of Technology in the West Bank, Karen Assaf has actively pursued research on water issues in the West Bank.

Naomi Chazan (Ph.D. Hebrew University) is a Knesset Member of the Meretz party, an associate professor of Political Science and the former chair of the Harry S. Truman Institute for the Advancement of Peace at the Hebrew University in Jerusalem. She is the author and editor of seven books and over fifty articles.

Souad Dajani (Ph.D. University of Toronto) teaches social and behavioral studies at Antioch College in Ohio. She is a former fellow in the program on Nonviolent Sanctions at Harvard's Center for International Affairs and the author of several articles about Palestinian women and the *intifada*.

Yvonne Deutsch is an Israeli feminist peace activist who has been involved in peace and anti-occupation activities since the 1970s. She was among the organizers of the Israeli Committee in Solidarity with Birzeit University, of the Committee Against the War in Lebanon, of *Women in Black* and of the *Women and Peace Coalition*. In her political work, Yvonne Duetsch has been committed to creating a political feminist culture of peace.

Nabila Espanioly is a feminist peace activist in Israel and one of the organizers of the Al-Tufula Pedagogical Center, in Nazareth – a women's center aimed at preschool education and at the empowerment of women. She has taught in the Women's Studies Program at Haifa University and has lectured widely on women in Israel, women in the Arab world, women and peace, and violence against women.

Afaf Ghazawneh has been a researcher for the Women's Studies Center in East Jerusalem since 1989.

CONTRIBUTORS

Suha Hindiyeh-Mani (Ph.D.) has been the director of the Women's Studies Center in East Jerusalem since its establishment in 1989 and has lectured in sociology at Bethlehem University and at the College of Art for Women.

Subhiyyeh Idris has carried out research for the Women's Studies Center, East Jerusalem and is now doing research for FAFO, the Norwegian Trade Unions in the West Bank.

Mariam Mar'i (Ph.D. Michigan State University) is the director general of the Educational Center of Early Childhood & Development of the Arab Child in Akka, Jaffa and the Triangle. She has lectured widely on Palestinian women in Israel and has been involved in Israeli/Palestinian dialogues.

Tamar Mayer (Ph.D. University of Wisconsin-Madison) teaches Geography and Women's Studies at Middlebury College in Vermont. She has been active in Palestinian/Israeli dialogues since the early 1980s. Her research focuses on gender(s) and nationalism.

Moshe Semyonov (Ph.D. SUNY Stony Brook) is the chair of the Sociology Department at Tel Aviv University, the co-author of *The Arab Minority in Israel's Economy: Patterns of Ethnic Inequality* (forthcoming) and of *Hewers of Wood and Drawers of Water: Non-Citizen Arabs in the Israeli Labor Market* (1987), and the author of more than thirty articles.

Simona Sharoni (Ph.D. George Mason University) is Assistant Professor of Peace and Conflict Resolution at American University and the author of *Gender and the Israeli–Palestinian Conflict: The Politics of Women's Resistance* (1994). She is an Israeli feminist and peace activist who has been involved in solidarity work with Palestinian women in the West Bank, the Gaza Strip and the USA.

Elise G. Young, a Middle East historian, is the author of *Keepers of the History: Women and the Israeli Palestinian Conflict* (1992). She has worked on Palestinian women and health under the Occupation and has lectured widely on the Israeli/Palestinian conflict.

ACKNOWLEDGMENTS

This book is going to print at a monumental juncture in the histories of both the state of Israel and the Palestinian nation: a time when their respective representatives have recognized each other's right to exist; a time when the celebration and the display of Palestinian national feelings are no longer illegal; a time when both Israelis and Palestinians are getting used to a new reality; a time when both people are learning to conceive of a new 'other'. At this historical juncture both peoples have a chance to claim the future of which both were, I believe, being robbed before September 13, 1993. Yet only time will tell whether these hopeful feelings are justified. By the time this book appears on the shelves of bookstores, the world will be better able to judge whether this act of mutual recognition has indeed yielded the promise of a new Middle East. Moreover, despite the excitement many of us are now feeling, the history of the last twenty-six years can never and should never be forgotten, as it has dramatically changed both Palestinian and Jewish societies forever.

The idea for this book has been with me for many years. It sprang indirectly from the Israeli/Palestinian dialogue groups in which I participated during the 1980s, where I learned to know better not only Palestinians but also myself; where I realized that only through dialogue and mutual understanding can both sides know and respect one another and work for change. And although I was convinced of the importance of dialogue before conceiving the idea for this volume in its present shape, working closely with many of the contributors has reaffirmed that conviction.

Many people have helped to make this project possible through their advice, suggestions, and encouragement. I am grateful to Suha Sabbagh, Joost Hiltermann, and Jonathan Price for helping me make the connections with some of the contributors; to Aleef Faransh, Beth Levison, Jennifer Nelson, and Marilyn Thorp for their help with interviewing, data analysis, library work, and with the transcription of more than 500 interviews; to the students in my Middlebury College Winter Term seminar 'Women in the Palestinian/Israeli conflict' for challenging me and for helping me sharpen

ACKNOWLEDGMENTS

my arguments; to Ann McLean for her wonderful secretarial aid; to Amanda Tate for her cartographic skills; and to Middlebury College for generously providing sabbatical time and financial support. In the process of finalizing the list of contributors and editing their work I was very fortunate to have the insights of Janine Clookey, Cynthia Enloe, Ayala Gabriel, and Joni Seager, who each in her own way helped me to focus on the important aspects of the project and who helped prevent me from becoming bogged down in paralyzing details. I am also grateful to have worked with Janice Monk, Janet Momsen and Tristan Palmer of Routledge. The close and immediate attention which the series editors brought to each of the chapters enabled me to deliver a stronger volume in a shorter time. Fran Breit, Julie D'Acci, Deborah Kutzko, Deb Lashman, Biddy Martin, Beth Mintz, Susan Murray, Susan Smith, Gabi Strauch, Carol Thayer, and Jane VanBuren cared for me regularly and provided encouragement whenever I needed it, from whatever distance. Their support, together with that of Ron Liebowitz, has made the editing and the writing process less lonely and more livable.

I want to give special thanks to four individuals. Although my parents, Artur and Shoshana Mayer, did not always fully understand my personal and political desire to bridge the Israeli/Palestinian gap, and often preferred that I be involved in non-political activities, they have nonetheless been encouraging and supportive throughout the project. Joanne Jacobson's enormous insights, constant criticism and editorial skills show throughout this book and were essential to my completion of it. I am greatly indebted to her for being so generous with her time, energy, skills and support. Finally, I could not have completed the book without the help of Tahl Mayer, who endured late dinners, my obsession with Palestinian and Israeli women's issues, and loud dinner-table discussions about issues in the Middle East; who often distracted me from my work with the needs of a five-year-old; and who ultimately kept me going. It is because of his view of the world, because of his excitement about a possible Palestinian state and about the potential for peace, that I am so optimistic. Because he and his generation will grow up with another reality we will have a better and, hopefully, more just Middle East.

Israel, the Gaza Strip, and the West Bank (September 1993)

1

WOMEN AND THE ISRAELI OCCUPATION

The context

Tamar Mayer

The brief and dramatic war that initiated Israel's occupation of the West Bank and the Gaza Strip in 1967 has turned out to be the source of long lasting changes – not only for Palestinians in the Occupied Territories, as indicated in much of the research on the region, but also for Israeli Palestinians[1] and for Israeli Jews. Israel's victory created an occupier/ occupied relationship which, much like the relationship between colonizer and colonized, placed Israeli Jews in the position of the powerful and the Palestinians of the Occupied Territories in the position of the powerless. This unequal power relationship which has lasted more than twenty-six of Israel's forty-five years of statehood has had significant social, political, economic, cultural, psychological and moral ramifications for both men and women, both Israeli and Palestinian. And although the hope of ending the Occupation is closer now than ever before, more than a quarter of a century of occupation has left a permanent mark on all these societies.

While the occupier/occupied power relationship denotes the dichotomous relationship between Israel and the Palestinians of the Occupied Territories, it leaves the close to one million Palestinians of Israel invisible. Although citizens of the state of Israel, most Israeli Palestinians identify nationally with their brothers and sisters across the Green Line, the Palestinians of the West Bank and the Gaza Strip who have been occupied and often oppressed by the very state of which the Palestinians of Israel are citizens. This frequently overlooked dilemma of conflicted identity remains an important indicator of the complexity of what it means to be either an Israeli or a Palestinian.

Although the Occupation has been accompanied by Israel's assertion of power over the occupied Palestinians, power has also been shifting in the relationship between Israel and the Palestinians of the Occupied Territories. It has in fact taken more than twenty-five years for Israel to begin realize how central to its own life the Occupation has become. Moreover, the personal and collective burdens of acting as either the occupier or the occupied no longer signal a clear and direct relationship to status, as they may have in the early years of the Occupation. As the Occupation continued, and especially since the beginning of the Palestinian *intifada*, the initially dominant occupier lost some of its power – and the initially powerless, the occupied, gained considerable strength. On one hand, the Occupation has provided an arena of resistance within which many Israeli Jewish women have found their political voice; while the imperative of resisting the Occupation has renewed forces of cultural, national and religious fundamentalism which have pressured Palestinian women to return to traditional roles. Yet, on the other hand, examinations of Jewish and Palestinian women's lives, as they unfold in this book, also show that the prolonged Occupation and its reinforcement of Israeli militarism have in many ways enhanced Jewish women's marginalization and changed Israeli society for the worse – while they have, at the same time, empowered the Palestinian women who have adjusted to new roles. Coping with the double or triple burden of occupation has given Palestinian women a political voice and strengthened their role in society as heads of households, as political and environmental activists, and as intermediaries between the Israeli army and Palestinian youth. Israel's surprise military victory of 1967, therefore, has turned out to contain a surprise of a different kind – that of subtly shifting power relations between Israel and Palestinian society under occupation.

HISTORICAL BACKGROUND

Although many of the chapters in this volume begin by briefly providing historical contexts for their specific theses, it is important to introduce at the outset the major historical players in this book: the Israelis; the Palestinians of Israel; and the Palestinians of the Occupied Territories. Because much of this information is available elsewhere this background will be short and limited in scope.

While the Israeli/Palestinian conflict has held the world's attention for only twenty-five or thirty years, this conflict is nearly 100 years old. Prior to the twentieth century, the Jews and Arabs of Palestine lived alongside one another in relative peace. But increased nationalism and, in turn, anti-Semitism in Europe renewed the appeal and sparked the reorganization of Zionism – the Jewish national movement – and Jews, in the early 1900s, began to seek a national homeland in Palestine. To these Zionist Jews, Palestine was the 'Zion' to which they had prayed for centuries and which had been the focus not only of their religion but of their culture as well. They generally perceived Palestine as empty and desolate, and gave little thought initially to the indigenous Arab population; later, when confronted by that population, they believed, or claimed to believe, that the land of Palestine could accommodate both peoples.

The last 100 years have seen massive Jewish immigration, first to Palestine and then to Israel. The migration of the early part of this century changed forever the pre-1882 Jewish/Arab balance in Palestine. Because Arab land was bought, leased and often expropriated by immigrating Jews, many Palestinians were pushed off the land, on which peasant culture and social organization directly depended, and forced to endure wrenching social and economic changes. Much of the 'colonizing'[2] of Palestine was accomplished by Jewish immigrants with the support of Britain, which had received a mandate over Palestine from the League of Nations in 1921. In its strengthening of a European presence in the region Jewish immigration evidently served British colonial needs; and British policies, especially in the 1920s, favored the immigration of Jews hoping to make the desert in their Biblical homeland 'bloom'. As in other British colonial adventures, the indigenous population of Palestine remained invisible to the British as well as to Jewish immigrants. But as Arabs' opposition to the Jewish migration and to British policies mounted, Britain reversed its immigration policies in the late 1930s and 1940s, limiting and at times prohibiting Jewish immigration.

Jewish claims to the land and British support for these Jewish claims constituted a clear challenge to Arab claims to the very same land of Palestine. Since for both nations the land of Palestine (later to become Israel) had been the only nationally acknowledged homeland, the Arabs of Palestine organized themselves to demonstrate and later revolt against these incursions.

The November 29, 1947 UN Partition Plan set the date for the end of the British mandate over Palestine for May 1948 and called for the partition of Palestine into separate Jewish and Arab states in order to satisfy both nations' claim to the land. Each of the states was to have full independence and sovereignty over its respective territory. But while the Jews of Palestine accepted the partition, the Arabs did not. They began in protest as early as December 1, 1947 to attack Jews and Jewish settlements, to loot and burn Jewish-owned stores in major cities like Jerusalem, and to attack residential neighborhoods in Tel Aviv. In early 1948 the Arabs of Palestine were aided by thousands of Arab volunteers from neighboring Arab states who organized loosely in scattered and often uncoordinated bands and raided Jewish settlements as the British presence in the region diminished. Although both Jews and Arabs attempted to set up governments, the Jews of Palestine were better organized because the Jewish Agency had already been functioning as a working government, while Palestine's Arabs had no comparable organization (Polk 1991). The military experience of these Arab bands was also inferior to that of the Jews. Vowing therefore to aid the mostly unorganized Arabs in Palestine in their war against the Jews following the UN Partition Plan, several neighboring Arab states, within which anti-Jewish and anti-Zionist sentiments were running high, pledged to intervene in the war between the Jews and the Arabs of Palestine. This they did, following British withdrawal, in May of 1948, after Israel had declared independence.

The exodus of Palestine's Arabs from major urban areas like Haifa, Jaffa, and Jerusalem and from rural areas, which began in December 1947 as

Jewish/Arab violence escalated following the UN Partition Plan, continued in waves until 1949 (Morris 1987). Many of the Arabs of Palestine, some of whom had fled willingly and some of whom had been forced to flee, found refuge in the West Bank and in the Gaza Strip as well as in Jordan and other Arab countries. These Arabs would later become known as the Palestinians. Many of these Arabs who settled in 1947 and early 1948 on the West Bank of the Jordan River and in the Gaza Strip faced in 1967 yet another confrontation with Israel, which caused many of them to flee again and become refugees for the second time in their lives. Thus discussions of the 'Palestinians of the Occupied Territories' refer to the Arabs of Palestine who have been under Israeli control since 1967, both those who settled in the West Bank and the Gaza Strip in the late 1940s and those who had been living there previously.

While most of Palestine's Arabs left for the Arab states (and some of these later found themselves under Israeli occupation), 20 per cent of them remained within the newly established state of Israel. Their numbers increased after the signing of the armistice agreement between Israel and Jordan, which placed several villages from Jordan under Israeli jurisdiction. These 'Arabs of Palestine' became then 'Arabs of Israel', as Israeli citizenship was granted to all minorities residing in Israel at the time. Despite the collapse of their social and economic infrastructures which resulted from the exodus of some 80 per cent of the Arab community and its economic and intellectual leadership, and despite their very weak political and military position, these Arabs were perceived by most Jews in Israel and by the Jewish leadership as a 'security' threat. Concerns with their pre-1948 activities and their connection with the neighboring Arab states, and many Jewish Israelis' fear that the Arabs of Israel would not identify with Israel and could therefore become a 'fifth column', led Israel to establish military rule which lasted until 1966 (Benziman and Mansour 1992) over Israeli Arabs, who only later would identify themselves as Palestinians. The military rule was intended to provide some basic services in civil matters and, at the same time, to address security issues in areas populated by 'minorities' and thus to provide an effective mechanism for controlling what was perceived to be an alien population (Benziman and Mansour 1992: 38). The military rule was maintained for this period of nearly eighteen years, even though the Arabs of Israel were not involved in subversive anti-Israel activity and were gainfully employed in the Jewish labor market. Since they were effectively separated not only from the Israeli population but also, as citizens of the state of Israel, from their families in the Arab countries, and since the military rule was lifted only about six months before the 1967 war, the Israeli Arabs had very little opportunity to define themselves as a non-threatening national minority in Israel. The 1967 war which would conclude with Israel's occupation of the West Bank and the Gaza Strip – centers of Palestinian population – therefore renewed contact between the two Palestinian populations and initiated the process of re-Palestinianization among Israeli Arabs.

WOMEN AND THE OCCUPATION: THE PROLIFERATION OF DIFFERENCE

Although the Occupation has affected men and women in many similar ways, men's and women's separate daily experiences have also caused them to experience the Occupation differently from one another. Because both Jewish and Palestinian women have typically been marginalized within their own societies it is important to understand the gendered impact that the Occupation has had on both groups of women. But because neither the Palestinians nor the Israelis are homogeneous, the Occupation has also been experienced differently according to class, ethnic or national group. Difference and conflict, even contradiction, therefore, paradoxically, are major common threads in this book which illuminates the impact of the Israeli military occupation on Jewish and Palestinian women. More specifically, the book suggests that more than twenty-six years of military occupation have exposed a set of often unarticulated internal conflicts for each of the three groups of women while, at the same time, also reinforcing existing loyalties among most of them. In addition, several of the chapters suggest that the prolonged Occupation has resembled a colonial relationship, even though military occupation is not by any means identical to colonialism.

Internal conflicts

One key result of the Israeli military occupation has been to expose previously unarticulated internal conflicts among Israeli Jewish women, and among Palestinians of both Israel and the Occupied Territories, as the Occupation has challenged social structures within all three of these societies.

One major conflict for Israeli Jewish women has emerged out of the dilemma of participating in 'national security' discourse, even while it is clear that maintaining the priority of 'security' prolongs the marginalization of women. Knesset Member, Naomi Chazan, in her dialogue with Mariam Mar'i (Chapter 2), Yvonne Deutsch (Chapter 5) and Simona Sharoni (Chapter 7) explain how the military, in which men participate more actively and for longer periods of time, has contributed to the organization of Israeli society; and all three of them assert that as long as Israel maintains its obsession with 'security', Jewish women will continue to be assigned a secondary status. Moreover, because the Occupation is intertwined for many Israelis with the discourse of 'security' (and for many Israelis holding on to the 'Territories' has clearly been an issue of 'security'), the ongoing

Occupation has, these authors believe, led to even further marginalization of women. For some Israelis, 'security' and 'occupation' have been conflated into one issue; for others, the Occupation has remained a security liability: for the longer Israel retains its hold on to the territories, the clearer it becomes that Israel cannot control the violence among Palestinian youth and the rise of Islamic fundamentalism which are forcing an even greater military involvement in the Occupied Territories. Yet, because men have retained such a central hold on security issues, Chazan suggests (Chapter 2) women have remained in any case marginalized in formal and organized politics as well.

Although many feminists in Israel, as Deutsch (Chapter 5) shows, are aware of how the national emphasis on 'security' marginalizes them and are critical of its impact on women, they often retain conflicting views about applying their own ideologically motivated critique to Israel's culture of 'security'. When their husbands and sons are called up for compulsory or reserve military duty or engage in 'security discourse' these women have difficulty separating Israel's security vision from their own existential fears, especially as their marginal role in the military further undermines their confidence in the legitimacy of their own judgment. The national myth and the language of 'security' have also concealed the extent to which the hostilities and aggression associated with 'security' affect Israeli women directly. Sharoni (Chapter 7) suggests that much of Israel's aggressiveness is turning inward, against Jewish women, rather than outward, against an external enemy, and that as a result violence against Jewish women has been on the increase. Moreover, the Hebrew language, Sharoni argues, reflects and reinforces the inequities of Israeli gender relations and cultural politics, and thus helps to cement the internalized perception among Israeli men that every woman is in essence an occupied territory.

Although Israel's obsession with 'security' predates the Occupation, it is only within the last two decades that feminism has given women tools for critiquing the power structure and for creating alternatives to it. While in the West, anti-war activities were carried out in the late 1960s and early 1970s, this critical consciousness was not part of Israel's life, mainly because wars in Israel were perceived unambiguously as a necessity for survival. But their disturbing inability to foresee how long Israel would hold on to the West Bank and the Gaza Strip, especially after the Likud Party's 1977 electoral victory, has motivated many Israeli women since the mid-1970s to organize against the Occupation. These activities have involved both Israeli Palestinian women and Jews and later also involved Palestinian women from the Occupied Territories.

While Palestinian and Jewish women's involvement in dialogue, demonstration, and extra-parliamentary political action to end the Occupation ultimately gave Israeli Jewish women, as Chazan and Deutsch both indicate (Chapters 2 and 5), a greater political voice and provided them entry into formal politics, it also exposed internal conflicts among and within these groups of women. As a result of these activities not only the similarities but also the differences between Jewish and Palestinian women of Israel became clearer. In particular, as Mar'i argues (Chapter 2), these joint activities have illuminated the complex position of Israeli Palestinian women who have found out that even in these women's forums their own needs are frequently not being addressed. Moreover, because the goals of these activities were to bring an end to Israeli occupation of the West Bank and the Gaza Strip, the needs of Palestinian women from the Occupied Territories have also generally taken priority, again leaving Israeli Palestinian women invisible. Thus the very forums that have provided Jewish women a political voice and an entry into formal politics have extended the silence of Israeli Palestinian women.

Palestinian women in Israel, as Nabila Espanioly (Chapter 6) suggests, also face a fundamental conflict between their civil identity as citizens of Israel and their national identity as members of the Palestinian people. While this conflict became virtually inevitable as a result of the 1948 war, the Occupation has since 1967 substantially sharpened it. Because 'administering' the Occupied Territories has consistently put Israel in a position of conflict with the Palestinian population there, Palestinian women of Israel have had to cope with the dilemma of being citizens of a state that oppresses the nation with which they identify and does not recognize their own needs as a national minority within Israel. Their involvement in extra-parliamentary political activity was therefore an expression of their effort to bridge this monumental internal conflict. The involvement in anti-Occupation activities which took many Israeli Palestinian women into the streets led them, in turn, to social and educational activism that empowered them within the public sphere and raised their political and social consciousness. At the same time, however, these activities have also made them face other conflicts. For because Palestinian women in Israel come from traditional patriarchal families in which women's participation in decision making is rarely possible, as Espanioly (Chapter 6) suggests, very few Palestinian women in Israel yet recognize the contradictions of the 'revolutionary' man who speaks publicly about equality and then expects his wife and sisters to defer to him and wait on him within the private sphere of home. In short, because Palestinian women in Israel are part of two worlds – that of the Westernizing one where they are being empowered by women's activity in the public sphere; and the world of traditional Palestinian society – they continue to be faced with conflict between their national and civic identities.

Among the Palestinian women in the Occupied Territories a slightly different conflict emerges: between social and national liberation. Many such conflicts that were previously ignored by Palestinian women in the Occupied Territories are now coming to the surface, as Dajani argues (Chapter 3), as women become more active in the national political arena.

Women's participation in the national struggle alongside Palestinian men, notably since the beginning of the *intifada*, has also raised these women's consciousness about the need to fight for social liberation, for it has brought to the surface frustration with women's traditional consignment to the private sphere. The social agenda of these women has become as important to many of them as their national agenda; and, determined to avoid the situation of women in Algeria whose participation in the national struggle did not yield social change, Palestinian women have worked to achieve both national and social liberation. But because Palestinian society remains largely traditional, even as it fights for national liberation, women are still finding themselves relegated within it to the private sphere, especially where 'motherhood' is concerned and especially as women have experienced backlash from Islamic fundamentalist forces.

Palestinian women of the Occupied Territories also face other internal conflicts. Because employment opportunities have been limited for Palestinian women and men in the West Bank and the Gaza Strip, as Semyonov explains in Chapter 8, many enter the market economy. Hindiyeh-Mani argues (Chapter 9) that entering the market economy as peddlers in the informal, *bastat* economy exposes a potential internal conflict for many of these women who have internalized the values of traditional Palestinian society. Their entry into the informal market economy simultaneously pushes the edge of their traditional role within the family and, at the same time, leaves room for the reinforcement of this role since these women only enter the *bastat* out of family economic necessity and they frequently turn over their earnings to male family members.

Thus the fights for national liberation and for social liberation have presented Palestinian women of both Israel and the Occupied Territories with an internal identity conflict, especially insofar as the heightening of 'Palestinian' nationalism has invoked the preservation of a shared past and a traditional way of life which have, historically, relegated Palestinian women to the private sphere and to a secondary status. At the same time, Jewish women's anti-occupation activities have also presented them with a conflict between national and social liberation in the form of conflict between their feminist and their national ideology. All these women, Chapters 2, 3, 5, 6, and 7 suggest, experience internal conflicts as they struggle, simultaneously, to oppose and to take part in events and discourse shaped by men who are not necessarily ready to listen to them.

Existing loyalties

While the Occupation has exposed social, national and ideological conflicts for both Palestinian and Jewish women, it has also reinforced existing loyalties within these groups. In working to end the Occupation, Israeli Jewish women have worked alongside Palestinian women from both Israel and the Occupied Territories, and as these women have become more involved in public political action they have all become targets of harassment whose sexual and nationalistic content has reinforced existing gender and national loyalties. These joint activities and these public incidents have also brought to the surface differences among Israeli Jewish women, Palestinian women in the Occupied Territories and Palestinian women in Israel, differences which remain serious.

In response to the Occupation's sanctioning of the separation of the Jewish and the Palestinian communities within Israel, Chazan argues (Chapter 2), women of both communities played important roles in crossing these barriers and sometimes in bridging them. These women were initially involved in dialogue among themselves and later engaged in dialogue with the Palestinian women of the Occupied Territories. Because women are not tied to the reigning paradigm and thus are freer to move and to challenge it, Chazan says, they communicate among themselves far better than if men are involved in these discussions. Gender loyalties, therefore, frequently cut across political loyalties for Jewish and Palestinian women. These meetings with other women increased Israeli Palestinian women's sense of commonly experienced oppression, inspiring them with the activist political example of Palestinian women in the West Bank and Gaza and of Israeli Jewish women and, therefore, highlighting gender bonds among all three groups and national loyalties among both sets of Palestinian women. Yet, while Israeli Palestinian women have been partners in dialogue with Jewish women and with Palestinian women from the Occupied Territories, this increased contact has also helped crystallize for them, as Espanioly argues (Chapter 6), their unique, and uniquely disempowered, social and political condition as Israeli citizens.

Moreover, many Israeli Jewish women who were involved in dialogues and in demonstrations against the Occupation found, as Deutsch suggests (Chapter 5), that they had also internalized state ideology about 'national security' and the importance of national 'self-defense' which reinforced existing national loyalties and violated developing gender loyalties. For despite their feminist critique of war and 'security', Israeli Jewish women tend in the end to join their men in accepting 'national security' as an important, almost sacred, notion which they cannot legitimately challenge.

Cross-gender national loyalties have also been experienced by Palestinian women as a direct result of the Occupation, even though their daily experiences are frequently different from men's. As Tamar Mayer argues in Chapter 4, Israeli military intrusions into villages, refugee camps, and homes have intensified Palestinian women's sense of national identity. And as Israeli soldiers have ruthlessly entered the Palestinian private sphere, disturbing family life and motherhood and violating traditional boundaries surrounding women's sexuality (during house searches and interrogations), women's anger towards the occupier has increased, further intensifying their national feelings. At the same time, men's daily experiences of oppression and humiliation at the hands of the Israelis, while different from women's, have also helped raise their national consciousness, so that the external force of occupation has reinforced internal national loyalties between Palestinian men and women.

Within the general context of the Palestinian campaign to achieve national autonomy on an economic level, Palestinian women have, as Elise Young (Chapter 11) explains, worked to address their own health needs and to develop their own health resources through organizations like the Women's Health Project. This imperative, again, has resulted directly from the Israeli military occupation of the West Bank and the Gaza Strip and its challenge to the integrity of Palestinian society. In signaling the recognition of the interconnectedness within Palestinian society between ill health and the subjugation of women, and between ill health and poverty, as Young argues, these local women's health organizations also reinforce a set of mutual obligations along both gender and national lines among Palestinians living under a military occupation.

Dajani (Chapter 3) also discusses the national loyalties that develop across class lines within the Palestinian women's committees. Although these women's committees were established in the late 1970s by young and educated women activists, their concern was not with the urban educated Palestinian women, but rather with working women and rural women. Further, by addressing women's daily concerns rather than international and political issues, the women's committees focused their fight for social liberation on the empowerment of local women and the creation of local leadership in rural areas, and indeed addressed a wide range of the Palestinian women regardless of class. It is this lesson the Israeli women peace activists have yet to learn, according to Chazan and Mar'i, who both agree that while it has addressed political issues, the Israeli peace movement has not created a grassroots leadership or adequately challenged existing class loyalties.

COLONIALISM AND OCCUPATION

Israel's occupation of the West Bank and the Gaza Strip was undertaken for military and security reasons rather than to promote Israel's economic development. But retaining military authority over these territories for a period that appeared, even as late as August 1993, indefinite has nonetheless clearly resulted in the strengthening of Israel's economic, military and even political control there. By asserting its presence in these areas Israel created, maintained and reinforced an unequal power relationship which closely resembles colonialism. Elise Young (Chapter 11), for example, argues that the treatment of Palestinian women relies on colonialist constructions of 'sex' and 'race' which permit the torture of Palestinian women political prisoners and the teargassing of women and nurseries, and limit women's access to health care. Even though, unlike occupation, colonialism is not carried out by a military force and is usually not enforced under conditions of territorial contiguity, many would argue that Israel has essentially colonized the Palestinian economy and politics of the West Bank and the Gaza Strip. Since colonialism is 'the establishment and maintenance of rule by a sovereign power over an alien people . . . [and the creation of] uneven economic development' (Johnston 1986: 59), we can say that indeed the Israeli military occupation seems to constitute a variant of colonialism, even though it did not introduce capitalism and its ideology to the Palestinians of the Territories.

For all practical purposes the social and economic situation in the West Bank and the Gaza Strip is similar to that of 'Third World' countries. In comparison to the Israeli economy, the Palestinian economy is weak and underdeveloped. Political restraints imposed by the Israeli government in order to purposefully create dependency on Israeli goods and thus limit Palestinian economic growth have also posed major obstacles to Palestinian development.

Israel has since 1967 restricted the quantity and the type of goods and raw materials which could be imported into and exported from the Occupied Territories. By limiting access to these resources, Israel isolated the industries in the Territories and caused their stagnation (Shadid 1988). In addition, because financial support for development ceased immediately after the Occupation in 1967, when Arab banks were forced to close and the Israeli bank branches which opened in their place limited loans to short-term and daily transfers only, there was little available capital for industrial investments. And because of restrictions posed by banking institutions across the border, notably in Jordan, the money supply to local industrial development dried up (Farris *et al.* 1993), further contributing to industrial underdevelopment in the Occupied Territories.

The stage of underdevelopment in the West Bank and the Gaza Strip is similar in many ways to what we find in 'developing' countries. Much of the Palestinian economy's dependency on Israel is clearly the result of Israel's occupation and of Israeli policies of, on the one hand, limiting economic growth in the Territories, and on the other, employing tens of thousands of Palestinians at below minimum wage in the Israeli labor market, which has simultaneously encouraged Israeli economic development and Palestinian underdevelopment. As many indigenous populations around the world are undergoing 'modernization' one would have expected the Palestinian economy to experience 'modernization' even if the Territories were not occupied by Israel. As in many 'developing' countries, the economic base of the Palestinian economy was agricultural, for it provided Jordan with most of its fruits, vegetables, and cereals (Brand 1988; Shadid 1988) and thus was of considerable economic importance to Jordan. But by, for example, limiting markets for the West Bank's produce, the Occupation has limited agricultural production there, 'freeing' many peasants for cheap labor in Israel, and in so doing accelerating the proletarianization which exacerbates women's marginalization.

Although in many countries undergoing 'modernization' women enter the labor market in increasing numbers, Palestinian women in the Occupied Territories, Semyonov (Chapter 8) argues, do not exhibit the same trend. By examining official Israeli statistics he concludes that while Palestinian men's rate of economic participation has increased since the early 1970s, the rate of Palestinian women's economic participation has actually decreased. But we know from examples around the world that not all workers' participation in the economy is recorded, especially if they are subsistence farmers, and that many who are involved in the informal economy remain unrecognized as productive workers. We can perhaps therefore assume that the decline in women's economic participation in the West Bank and the Gaza Strip may be at least in part the result of their unrecorded participation in the informal economy, especially because Semyonov's data was compiled from official records on the formal sector.

As the economic situation in the Occupied Territories has worsened since 1967 and many Palestinians have become involved in the informal economy, peddling, or *bastat*, has become a well known phenomenon among Palestinians, as Hindiyeh-Mani, Ghazawneh and Idris argue in Chapter 9. While there is still debate about the role of informal activity in economic growth and integration into the capitalist mode of production (Weiss 1988; Hosier 1987; Connolly 1985), this study of women peddlers suggests that Palestinian women's position in production is becoming more integrated into capitalist production and that *bastat* work provides some of these women with the opportunity to accumulate small capital and thus to enter into commodity production or even capitalist production. But, at the same time, because women's *bastat* work is part of the informal sector and these workers come from traditional patriarchal families, women remain dependent on men within the patriarchal structure and thus their status as free wage laborers, inherent to capitalism, remains questionable.

The colonial economic relationship with the Occupied Territories in which Israel has engaged during the last two-and-a-half decades has not been confined exclusively to the economic sphere, but has also been carried out on the environmental level. As the growing literature on the relationship between (neo) colonialism, or Western modernization, and the environment suggests, environmental concerns frequently yield to economic growth, for when increased production becomes an important priority, environmental and human health issues usually take a back seat. Because of the Israeli military occupation, Palestinians' authority over their own land has been limited, contributing, as Assaf argues (Chapter 10), to the exacerbation of environmental hazards. The Israeli military presence in the Occupied Territories and the establishment of dozens of Jewish settlements there, which use a disproportionately large share of the water available in the West Bank, have severely reduced Palestinians' access to water and lengthened women's journeys to the well. Moreover, because Palestinians do not have regional or even local plans for the disposal of solid and sewage waste, and because farmers often use chemicals as fertilizers, open channels of raw sewage and chemicals often wash into the soil, contaminating the aquifer and polluting the available water resources. And because women spend more time in their own communities than men do, they are more exposed to such contamination and suffer its environmental consequences more often.

The situation in the West Bank very much resembles the case of many 'developing' countries in which, as part of the modernization process, developers have dumped and encouraged the use by local farmers of chemicals manufactured (and often banned) in the West. While Israel has sought to reduce contamination from herbicides and pesticides used by Israeli farmers, it has not done so for the Palestinian farmers of the Occupied Territories. Moreover, because many Palestinian farmers believe that use of these poisonous chemicals is a sign of advancement they have used them freely, as Assaf argues, contaminating their soil and produce and endangering the lives of many Palestinians, especially women at the workplace and at home.

But the environmental impact of the Israeli Occupation has generally been even more severe than that of colonization, or even modernization, as experienced around the globe. While during the development process deforestation, for example, is regularly practiced either as an economic activity, for profit, or as a way to allow for future economic projects, deforestation under military occupation is carried out either randomly or for tactical reasons. It is well known that 'militaries are major environmental abusers . . . and privileged environmental vandals' (Seager 1993: 14), and in the case of Israel's Occupied Territories the military has regularly practiced environmental destruction as a matter of strategic policy under, as Assaf claims, the pretext of security measures. Deforestation of Palestinian land has, therefore, been carried out in order to create a road infrastructure to newly developed Jewish settlements, as a means of disrupting Palestinian reliance on their fruit trees, and as a mechanism for controlling Palestinian youths who hid in the groves during the *intifada*, waiting to throw molotov cocktails at passing cars. In addition to the environmental destruction which has been carried out in the context of intentional policy during the *intifada*, individual Israeli soldiers acting apparently in what they interpreted as the spirit of military occupation – as opposed to colonizers – attempted to contaminate Palestinian water tanks in urban areas with urine, or shot at water tanks, hoping to limit the water supply available to occupied Palestinians.

More than twenty-six years of occupation have changed both Jewish and Palestinian societies in innumerable ways. While initially the Occupation may have separated these societies, it has also brought them together. If it were not for the Occupation neither the mutual awareness nor the internal loyalties of all three societies – Jewish Israeli, Palestinians in Israel, and Palestinians of the Occupied Territories – might have come to the surface. Equally, without the Occupation the internal conflicts in all these societies might never have been exposed and challenged. Thus, perhaps paradoxically, the Israeli military occupation of the West Bank and the Gaza Strip seems ultimately to have led Palestinian and Jewish women to challenge their social, political and economic situation and to strive for change.

NOTES

1 The term 'Israeli Palestinians' refers to the Arabs of Palestine who remained within the newly established state of Israel in 1948. They used to be identified as 'Israeli Arabs', a term coined by Israel to denote their national identity, on the one hand, and their Israeli citizenship, on the other. In this book they will be referred to as Israeli Arabs, as Israeli Palestinians, or as Palestinians in Israel. Each of these terms carries with it a defined political and ideological position on the relative importance of the national or the state identity of these women.
2 Since the early years of the conflict over Palestine, the indigenous population has seen the Jews as colonizers and believed that Zionism was motivated by imperial and colonial interests. But Zionism, the Jewish national movement, is far from being a colonial movement, mainly because the national elements are well embedded in Jewish religion and culture and because Jews settled Palestine in order to return 'home' and not in order to develop economically. For a detailed discussion see M. Benvenisti, *The Sling and the Club*, 1988, pp. 14–29.

BIBLIOGRAPHY

Benvenisti, M. (1988) *The Sling and the Club: Territories, Jews and Arabs*, Jerusalem: Keter Publishers (in Hebrew).

Benziman, U. and Mansour, A. (1992) *Subtenants: Israeli Arabs, Their Status and the Policy towards Them*, Jerusalem: Keter Publishers (in Hebrew).

Brand, L. (1988) *Palestinians in the Arab World: Institution Building and the Search for State*, New York: Columbia University Press.

Connolly, P. (1985) 'The politics of the informal sector: a critique', in N. Redclift and E. Minigione (eds) *Beyond Employment: Household, Gender and Subsistence*, Oxford and New York: Basil Blackwell, 55–91.

Farris, A., Fishelson, G., Jubran, R., and Nathanson, R. (1993) 'The labor market in the territories', Discussion Paper, Tel-Aviv: Histadrut-General Federation of Labor in Israel, Institute for Economic and Social Research.

Hosier, R. (1987) 'The informal sector in Kenya; spatial variation and development alternatives', *The Journal of Developing Areas*, 21, 383–402.

Johnston, R. (1986) *The Dictionary of Human Geography*, Second Edition, London: Basil Blackwell.

Khalidi, R. (1988) 'Economy of the Palestinian Arabs in Israel', in G. T. Abed (ed.) *The Palestinian Economy: Studies in Development Under Prolonged Occupation*, London and New York: Routledge, 37–70.

Mansour, A. (1988) 'The West Bank economy: 1948–1984', in G. T. Abed (ed.) *The Palestinian Economy: Studies in Development under Prolonged Occupation*, London and New York: Routledge, 121–138.

Morris, B. (1987) *The Birth of the Palestinian Refugee Problem, 1947–1949*, Cambridge: Cambridge University Press.

Polk, W. (1991) *The Arab World Today*, Cambridge: Harvard University Press.

Seager, J. (1993) *Earth Follies: Coming to Feminist Terms with the Global Environmental Crisis*, New York: Routledge.

Shadid, M. (1988) 'Israeli Policy towards Economic Development in the West Bank and Gaza', in G. T. Abed (ed.) *The Palestinian Economy: Studies in Development under Prolonged Occupation*, London and New York: Routledge, 121–138.

Weiss, D. (1988) 'The informal sector in metropolitan areas of developing countries', *Economics*, 38, 60–73.

2

WHAT HAS THE OCCUPATION DONE TO PALESTINIAN AND ISRAELI WOMEN?

A dialogue between
Naomi Chazan and *Mariam Mar'i*

Both Dr Mariam Mar'i, the General Director of the Organization for Akka women and Professor and Knesset Member, Naomi Chazan, have played important roles in protesting against Israel's occupation of the West Bank and the Gaza Strip since the early 1970s, well before anti-occupation activity became popular. Their political and social activism has often taken the form of a Palestinian/Jewish dialogue and has touched in important ways the lives of Jewish and Palestinian women living both in Israel and across the Green Line. Therefore I asked these two prominent women scholar-activists to extend that dialogue here. They agreed to meet once again and, more specifically, to assess how women in their respective communities have been affected by the Israeli military occupation. In early July 1992, about ten days after Israeli elections in which the Labor party returned to power following fifteen years of Likud's right-wing government, I facilitated this discussion between Chazan, at that time, newly elected to the Knesset,[1] and Mar'i.[2]

Despite my presence, and that of the tape recorder, their exchange was candid and illuminated a number of key issues, including the similarities between Israeli military rule (over the Israeli Arab population between 1948–1966) and Israeli military occupation (over the Palestinians in the West Bank and the Gaza Strip, since 1967); the impact of national security on the lives of women in Israel; and the crucial role of dialogue. Equally important, the model of 'dialogue' which Chazan and Mar'i have initiated here helps, from the start, to confirm the agenda for this entire book. In that spirit, I hope that this volume will contribute to building the self-awareness, the knowledge about one another's communities, and the tolerance and mutual respect on which the achievement of a peaceful solution to the Israeli/Palestinian conflict ultimately depends.

(Tamar Mayer)

PALESTINIAN AND ISRAELI WOMEN

NAOMI CHAZAN: What has the Occupation done to Palestinian and Israeli women? Maybe I'll start by answering for Jewish women, and we'll see where we go from there. Because in many respects, without our knowing it, the Occupation was the first stage in setting Israeli women back politically. It was not just 'the Occupation'; it was the Occupation and '67, which heavily militarized Israeli society and essentially made it more apparent that women were back-seat drivers in the business of running the Israeli state. Anything that had to do with women was marginalized – and, much more important, there was the expectation that women would support whatever men did. Over the last ten years that expectation has had a second effect, of political liberation, for Israeli women, because it has also made women feel that if they are not part of this system then they have much more freedom to work against it or to change it. And I think that the fact that more and more women have gone into protest movements and that more women than otherwise would have done so have gone into formal politics is an indication of a sense of liberation. But ultimately I think that Israeli Jewish women have been occupied by the Occupation in the true sense of the word. And that we are almost having to use an internal, feminine *intifada*, a Jewish women's *intifada*, in order to free ourselves from the Occupation. We Israeli Jews almost need to get rid of the Occupation in order to liberate ourselves, and women will be, I think, among the first to do that in the real sense of the word.

MARIAM MAR'I: I agree with you. I believe that women understand most fully what oppression means, because they live it and thus are aware of it. Liberation has to come hand in hand with personal awareness and with deep understanding of oneself either as an occupier or as oppressed. I do believe that the oppressed (whether Palestinians or Jewish women) who understand oppression and have openness, responsibility, and commitment can free themselves and their oppressors from oppression. Assigning this responsibility to women may seem like blaming the victim, but it really isn't. Freeing oneself and one's oppressor from oppression is a totally different process.

But to go back to the issue of the Occupation and its effect on women. It's interesting, Naomi, that you said that the first reaction of Israeli Jewish women to the Occupation set Jewish women back, put them in the back seat. For during the same period, Palestinian women in Israel were probably taking their first steps toward understanding or becoming more aware of their role as women. And don't forget that this almost coincided with the lifting of the military rule. The military rule affected the whole community. Nevertheless, it has had its own specific impact on women, who are more sensitive to issues (I say that with pride, not as a negative observation). I don't want to go into history; but for us that was, on the one hand, a starting point of one stage, while on the other hand we got the military occupation of 1967. The Occupation of 1967 reminds me of what I went through personally when I gave birth to my daughter Rasha, my second child. I had my first child with an epidural injection, in the US. I really did not feel a thing, and so I didn't know what it meant to give birth in the natural way. So here she was, my second one, born in Israel the natural way, only because nobody allowed me to have her the easy way. I waited at home in labor for a very long time. When I finally got to the hospital it was at the last minute, and within a few minutes Rasha was born. I could not believe it, and I kept on asking the nurses what had just happened. Then, while I was in euphoria, a lady on the other side of the divider went into labor, and I could hear everything: the screaming; the crying; the writhing in pain. And every word, every screech, and every cry I felt again and again. It was like going through labor all over again. Really. Damn it, I thought I had just done that already.

Similarly, the occupation of 1967 reminds me of 1948: an experience that I just grew up with; like something that I already knew, something that was just there. I didn't go through the pains of military rule personally, because I was too young, a child at that point. Yet all of a sudden I had to re-live that rule through the occupation of the West Bank and Gaza Strip. I'm speaking here on a circumscribed, personal level, but basically what I'm trying to say is true also on the macro level. Israeli Palestinian women sympathized with their Palestinian sisters and brothers in the Occupied Territories; they really watched because they saw the suffering of the others. Israeli Palestinian men had other problems: Palestinian workers from the Occupied Territories came and took their work places. So Israeli Palestinian men had very conflicted attitudes towards the situation of their occupied brothers and sisters in the West Bank and Gaza Strip. But that was not the case for Israeli Palestinian women. On the contrary. I'm not saying that their political awareness in those ten years simply followed Israeli women's, and that when Jewish women started to become more assertive so did Israeli Palestinian women. But their own reactions coincided with those of Israeli Jewish women at that stage, even though they came from a very different place. Gradually, through developing more maturity and more confidence in their own abilities, Israeli Palestinian women have gained through the years on a number of different levels, and thus the Occupation has strengthened the political involvement of Israeli Palestinian women. I don't want to be misunderstood here: this involvement has not reached the desired level, and not all women have been politically active. But there have been improvements. And there is no doubt that the Occupation and everything around it have given legitimacy to political activity on the part of Israeli Palestinian women. Because this is a noble cause. These political acts are not committed because women want to achieve more freedom for themselves, to dress more Western, or to gain personally – and the fact that it is larger gives their cause a push.

CHAZAN: I want to pick up on two of the many interesting things that you said. I think that the analogy of feeling the pain of the other woman's difficult labor after the ease of your own daughter's birth is important. Your pain was a second-hand pain, not a first-hand one. Isn't this, in essence, the difference between Palestinian Israelis and the Palestinians living outside the Green Line?

MAR'I: Yes, no doubt about it. But this does not mean that second-hand pain is not painful. Sometimes it's even more painful, especially when your whole body is still sore and you are trying to heal it – and then all of a sudden you experience the pain again. Although you want to forget it, you can't. You've passed through your own pain in a gradual manner with many disappointments, slowly accepting it, and now you experience the other person's pain at once, without any pre-pains. It's a shock. My body had not yet recovered when I heard my neighbor screaming in pain, and it was very painful for me to listen to her; so much so that I could not distinguish which was my own pain.

CHAZAN: I want to pick up a second point very quickly, though I don't quite know what to do with it. As an Israeli woman, I think that some women have never given birth at all, either directly or indirectly. It's as though they took a baby and cared for it without any of the pains. I think these women are the ones who to this very day support the Occupation. I think others took the baby, brought it up, and then asked: where did this baby come from? And that's where the political awakening, at least among women doves, began.

MAR'I: The second-hand pain issue turns out to be a very interesting one. You're trying to differentiate precisely between my pain and that of our sisters in the Occupied Territories. You even help to sharpen boundaries. Yet the differences between how we feel and how our real dove partners among the Israelis (those whom we think of as our real partners) feel is a totally different matter. Maybe because, as you said, these Israelis took a baby, brought it up, and then asked where it came from. Even though we and our partners seem to agree on so many things, it seems that there are still many things that have to be explained to them, things that seem so obvious to us, things that we take for granted. It always surprises and shocks me that women on the Left, activists against the Occupation who share with us so much, have to have explained the simplest things. But maybe one should not be shocked, after all; and maybe your analogy can ease my anger and increase my understanding of these kinds of situations when they arise.

CHAZAN: So what do you do with Israeli women, whom I just described as virtual virgins?

MAR'I: I don't know. That's why I'm saying that you sharpened the distinctions among these different levels of women's experiences. I'm just now beginning to realize both how absolutely right your classification is and, at the same time, how unfair these distinctions are to people like me. I hope I can make myself clear here. I think it is unfair that people sometimes demand that we understand and take into consideration things which they themselves haven't gone through and do not understand.

We have to think of different categories – our virgin sisters; our watcher sisters; our second-hand experienced sisters as distinct from our directly experienced sisters – which need to be clarified and addressed by those of us who really believe that we have a role to play. I'm not saying that we can be totally convincing with this analogy; but at least if we refine it we may be able to use it to get our point across more clearly, since we really believe in dialogue not just for the sake of dialogue but for the sake of changing attitudes and arriving at actions.

CHAZAN: This dialogue is also for the sake of understanding ourselves better. For me, one of the reasons for dialogue is to sharpen the differences between us, because if you understand the differences you understand yourself better, and if in the process there is growth in self-awareness, that in itself has a value. For if it brings self-awareness, dialogue can lead to action that can eventually lead to more understanding.

MAR'I: I agree with you one hundred per cent. A saying of Christ's which I love very much is that good deeds start at home. If we can't understand ourselves and be self-aware, I don't believe we can get our own points across to anybody. When people ask me if I play sports, I say that I don't play sports except ping-pong. If I find a good partner, I'll play him or her. I can pass a few balls, but he/she has to be very good to enable me to draw out the little good that I have inside me. Similarly, dialogue without a really good partner is meaningless to me. It doesn't add much, and I feel that I am circling around myself. That is why dialogue with Naomi has been so essential for me, and that's why I love to think out loud among smart people from varied backgrounds and cultures – because it definitely helps sharpen the discussion.

CHAZAN: I just want to comment on that last point with respect to another thing that the Occupation did. I can say for myself that I was too young before '67 to even think about it; but since '67 it's been a very long haul to try and reach any inter-communal communication in this country. The truth is that the Occupation has sanctioned the separation of the communities. And women have had to play a very, very important role in trying to cross these barriers, though they have not always done so successfully. Because, as we both know, the Occupation has prevented people from mixing; we have all lost a lot of our own identities. Because without the contact there hasn't been the creative tension that makes each community grow.

MAR'I: Look at Jewish history. Although the history of Jews in the Diaspora is at many points a painful history for you, many of us have sometimes envied it. This is why there are such mixed feelings about our own Diaspora, about the Occupation, and about all the damned suffering. We feel that the unique achievements of the Jews in the humanities, the sciences, philosophy, and psychology are partially a result of their pain: of being oppressed, of being chased. Although sometimes it's been blown up and exaggerated, we really identify with that entire experience; I think that we see the advantages of it. Deep down, unconsciously – and sometimes we even say it among ourselves, in secret – we wish that the Occupation could last a little bit longer. Not because we are a self-hating people – but because we see the power that can come from pain. That's why I say that the Occupation has been so controversial and so hard. And that's why I say that if anybody can understand this, it's really the Jews. That's why we see ourselves as the wandering Jews.

The negative aspects of the Occupation – which are numerous – have all been openly stated; and under such circumstances people tend, justifiably, to overlook their knowledge of other effects, and that is what makes this discussion so sensitive. I do not want to sound as though I think that the Israeli government is doing us favors but, instead, to emphasize that there are always two sides to every coin. As one of my professors said, there is a lot of beauty in and around a garbage bin, even though it remains a garbage bin. As a result of the Occupation a lot has happened in the Palestinian community: the assertiveness; the maturity; the ability to stand up for one's rights. So whether we like it or not, occupation has had a liberalizing element for the self-aware. The only way for that liberalization to succeed, ultimately, will be through praxis.

TAMAR MAYER: And is that liberalization something that is unique to the Palestinians in the Territories, or has it been equally applicable to Israeli Palestinians? I think that their oppression under the Israelis has been different.

MAR'I: Yes, very different, because the presence of the military has been much much more intense in the lives of the Palestinians of the Occupied Territories. The Israeli Palestinians' experience was very different: the effect of Israeli rule was more gradual, although it was deep and it killed us at the first stage. Palestinian Israelis have not experienced any military confrontation. We have experienced segregation and ghettoization, but we have not had to face or to live with the military. I lived in the Old City of Akka[3] throughout my youth without encountering any real violence or any demonstration for a cause. The military rule was not as intense for us as it has been for the Palestinians in the Occupied Territories. We did not have our own *intifada*. It was totally different.

CHAZAN: On the one hand, I understand what you are saying. On the other hand, I do have to question it. I want to be careful not to glorify discrimination and suffering. There is a danger in trying to draw out the one potentially good thing about occupation or oppression or suppression, and magnifying it – and forgetting that the reality is very different from that. The reality actually is the opposite.

MAYER: It's just that I think that there is beauty in garbage, and after you sit for a while in a terrible situation you have to learn how to get some good out of it. That does not mean that the situation should continue. On the contrary.

CHAZAN: There may be beauty in garbage, but, to continue that comparison, garbage is garbage, and garbage is unhealthy. Another point, more from the Jewish Israeli side, is that recognizing a certain situation doesn't mean that you will avoid it or prevent it from happening to somebody else. It means that one gets so involved in one's own suffering that when one sees someone else's suffering one almost rejects any comparisons. This is the classic, or what has become the classic, argument about the Holocaust. Nothing was as bad as the Holocaust, and therefore anything – anything – that Israel does in the name of national security to anybody else, especially the Palestinians, is not comparable. Therefore in a sense the defensive mode, that you can't (I think) even begin to understand, separates us. Being systematically discriminated against makes you more self-aware and makes you value freedom more; but it doesn't make you any more tolerant, any more understanding, or any less oppressive in certain situations than those who have oppressed you.

MAR'I: No one except those who directly experienced the Holocaust can shed light for us on this. We have people who have suffered greatly, yet they have come out more human than any of us who did not experience their pain or their agony. I doubt that this is a matter of intelligence, as I have seen also very smart people who come out of the Holocaust experience rather vengeful. I'd really like to know what components of character enable some people to experience inhumane treatment and come out humane, while others come out cruel. Maybe we have to study this phenomenon.

CHAZAN: I think that we have at least these two types: the very humane; and those for whom the order of priorities has been self-preservation and self-protection above anything else. On one hand, if self-preservation or survival becomes a value, then no other values can be entertained. On the other hand, I agree with you: exactly the opposite effect can occur as well, when only circumstantial values have to be protected and everything else becomes secondary. Therefore preservation in the broader sense cannot be obtained unless certain values are pursued very systematically. And I think that there are a lot of shades between these two extremes. Almost any people has a crisis when it gets out of a period of severe oppression or discrimination, and in many respects Jewish Israel is still in that crisis because we haven't resolved it for ourselves: instead of even discussing it among ourselves, we have separated it into different camps. Each camp has its own definition of the issues, and there is no discourse on these problems. There is just assertion, and that's why Israeli politics looks the way it does.

MAYER: Do you think that there is any difference between women's and men's experiences of these issues? Do you think that dialogue occurs among women which doesn't occur among men?

CHAZAN: Let me separate these two things, because I really think that sometimes by comparing men and women we are not doing either women or men full justice. In seeking to make such clear distinctions we are not able to express the variety of either women's or men's responses in situations where the ranges may be slightly different or even sometimes very different. The Jewish woman's experience in Israel is, first and foremost, varied. There is no single cast. Among Mizrahi women experiencing their third generation of poverty, for example, you can get the following responses to the Occupation: a right-wing extremist response; a moderate response; or an indifferent response. You can also find these women among the pathbreakers to peace, for they are the only Israeli women who are going to understand Palestinian Israelis or Palestinians. So I have problems generalizing on that level at all.

I think that the second part of your question is a quite different one: do women discuss things among themselves better? In some sense, though it contradicts everything I just said, I would say *yes*, mainly because, again, we're less tied to the reigning paradigm. This gives us much more freedom of movement, and in searching for answers we may be willing to go farther afield than some of the men. We are not held back. Also, we admit we are confused. I think that just the ability to articulate the fact that you don't have all the answers and that there are contradictory things going on is a breakthrough. When you're looking for answers, you'll initiate conversations and encounters that you previously did not because you were afraid to reveal your own vulnerabilities. Whereas if you rely on existing wisdom you remain tied to what was previously considered taboo.

MAYER: And do you think that this need to question, the admission of ignorance, is connected in some way to the situation of controlling another people and to the Occupation itself? Or do you think that such questioning has just been a natural feature of the growth process in Israeli society and that maybe the Occupation itself is unimportant in this kind of a discussion?

CHAZAN: The Occupation can't be unimportant, because it is part of the particular trajectory of Israel, of *this* society, and therefore can't be separated out. I think that the Occupation has played an integral part in Israel's growth. I'm now going back to the beginning of our discussion, in an odd kind of way, to my statement that the Occupation enslaved Israel.

MAYER: What do you mean, 'enslaved Israel'?

CHAZAN: We have now reached the point where we need to go through a second liberation. Before '67 we could make our own mistakes, live with them, and correct them; but since '67 that has changed. Now the '67 war was an historical event (though here Mariam may disagree with me). There are still many different interpretations of this event; none of us has yet really succeeded in grappling with the war. But it is not the '67 war *per se* that interests me. What interests me is the fact that we have held on to the Occupied Territories and have continued the Occupation. The tragedy of '67 is, I think, that it was a necessary war from the point of view of maintaining Israel's survival, and yet it enslaved Israel because we didn't realize the monumental implications – for ourselves, let alone for the Palestinians – of holding on to the Territories. The Occupation enslaved us because it tied Israel's destiny to the destiny of the Palestinians, and I think that it has taken us virtually a quarter of a century to realize how intertwined Israelis and Palestinians became.

MAR'I: We became twins.

CHAZAN: Yes, we became twins. And the fact is that Israelis can't really be Israeli Jews and Israel cannot really be an independent state, because the Occupation has made us dependent on what happens in the Occupied Territories. If there is only one reason – totally selfish from the Israeli point of view – that the Occupation *has* to be terminated, it is that otherwise we will never free ourselves from this situation. Never. That's why I'm saying that in a sense we're facing our second decolonization. Our people's colonization was of our own making: we colonized ourselves in 1967.

MAR'I: This colonization is much stronger, much more dangerous than an external one because under conditions of internal colonization you are not in a situation which is concrete, where you can see and fight, where you can face your challenge and react. I think that at this critical stage Israel is very lucky – probably as a result of the system itself – that it still pampers its own thinkers and allows them to speak out. Its thinkers can be the conscience of this state. The mere fact that these voices exist is one of the good things if not the best thing that has happened through these last twenty-five years of occupation, because the internal colonization is so subtle; all of a sudden you wake up and it's there and it's so hard to free yourself from it. While these thinkers on the Left have frequently been called self-hating Jews, I think that they are the most loving Jews. Their desires actually coincide with my own needs as a Palestinian and show me that my partner is a humanist. The fact that the Left exists in Israel and is constantly working to revive or awaken the conscience of the state also counterbalances the results of conscious or unconscious values and behaviors which are absolutely there. Their counterbalancing act shows that many on the Left still have some hope for a better Israel, for a long-term Israel.

MAYER: I appreciate the whole issue of the Left as the Jewish conscience, on a certain level – but what do we do, then, with the people on the Left who are clearly part of the army, who obey orders despite being conflicted?

MAR'I: The mere fact that one does not obey orders blindly is important. At one point in my life I didn't understand, or want to understand, that. I could not understand how a person who really believed in the values of the Left could accept the fact that his or her son would be drafted by the Israeli army. Again I want to go back to the analogy of the second-hand experience. At one point I believed that this behavior represented a double standard, and I took it personally; not as Mariam, but as a Palestinian. But the more I mature the more I become aware of my own conflicts and even of my own double standards on issues. I know that life is not black and white; it is much more complicated than that. My hope for this conflict lies with the questioning, with those who do not accept orders blindly. This hope is something that we all have to hold on to, until questioning is no longer only questioning and has become action. This questioning has to go through a process which I understand better now than ever before because I (as well as Palestinian society) am also growing and maturing and getting closer to understanding what I will and will not do.

MAYER: Let me interrupt just to ask for clarification. When you talk about 'the Palestinians' here, do you mean the Palestinians in the Occupied Territories, the Israeli Palestinians, or both?

MAR'I: Both. The partnership between Palestinians and the Israeli Jewish Left used to be very naive. We constantly felt disappointed and even betrayed by the Israeli Left, and they felt the same way about us. But there is no doubt that when Yossi Sarid told the Palestinians to 'get lost,'[4] it was a turning point in the relationship between Palestinians and Jews. I have to give him credit for that. Because of that shock, many dialogues occurred. The first reaction was disappointment and anger. But the second reaction was admitting: 'Well, let's face it, haven't we really taken things too much for granted? Let's reevaluate things and put them in perspective.' I think that the relationship today between the Israeli Jewish Left and the Palestinians has become more tense, maybe because of Saddam; yet it is much healthier and much more mature than it used to be.

CHAZAN: I absolutely agree with you on this point. The first time I understood the real value of discussions with Palestinians was when I met the first Palestinian whom I really didn't like. I didn't like him personally, I found him abrasive; I didn't agree with him. And I said to myself, 'Wow, that's not possible, because you are supposed to like all Palestinians.' That's when I realized that these discussions were exactly what I needed, because as a result of them I was starting to make distinctions, and that had a humanizing effect. The real humanizing task is not constantly to quest for understanding, but rather to accept the fact that just as you do not like certain Jews you may also not like all Palestinians – and there is nothing the matter with that, it's totally human. In a sense what you are saying about Yossi Sarid is exactly the same thing, although you know I disagreed with his statement and I thought it was mis-timed and mis-placed. When I met this particular Palestinian whom I didn't like (and I still don't like him), it also forced me to think not about Palestinians but, instead, about my own reactions. Because we've been at this business for a long time, you know very well that many of the 'dialogues' (I don't really like the term) at the beginning weren't dialogues at all: each side told the other side what they should do, but neither side ever talked about themselves. And now we've come a long way, because we talk about ourselves openly to each other; and I think that is the real change between stage 1 and stage 2. That's point number one: it's this capacity to talk to you and understand myself better that is so important.

Secondly, this sharpens for me something else that I want to discuss – and that's the dilemma of many Israelis like myself, for whom there is too great a gap between our identities as Israeli Jews (and I make no bones about my identity as an Israeli Jew) and our political opinions and political values; between what we would like to see and our frame of reference in terms of national identification. We can't reconcile this gap because, on the one hand, we feel identified and, on the other hand, we disagree with what we identify with. The only good thing that can come out of this kind of tension – which, by the way, is not a manufactured tension; it's a real tension that I have felt for some time – is the possibility of translating it into precisely what you said, into some kind of action. Because nobody is going to close that gap for us. We are the ones who are going to have to close the gap, by changing a situation where it is easy for us to reconcile these two things: what we think should be; and our desire to identify with something that over the years we have become more and more alienated from. Now some people within the Israeli Left have solved the problem by simply not identifying, but they are also ruthless.

MAR'I: This is, I think, where my little understanding of the other came from. Those Israelis who became alienated became disaffected, and then they became marginalized – and then the question for me as a Palestinian became: who are my real partners? In order to have your perception of life actualized or realized, you need partners.

Here I cannot be completely frank, because I am stripping myself naked and I risk betraying my friends. What I'm saying about the different parts of the Israeli Left and about other groups within which I personally locate myself as a Palestinian is that we are aiming at healthy interactions that will bring us to a real partnership. For us, too, it goes step by step. With every step we take we make another compromise and gain more confidence in ourselves. You cannot compromise if you are not confident, if you are not secure, and if you are not mature. And this is how you get closer to bridging the gaps. This is also true for the other side. If on the other side there is no maturity and no self-confidence, compromise cannot occur, and then – because these interactions are very much dialectical and reciprocal – things get stuck somewhere, they don't move or they take much longer. There have been stages where it took forever to move one inch; while at other times you feel like you are running and you need to catch your breath, to collect your thoughts.

I really do believe that we are now facing a new era. These past elections have suddenly re-energized me. I'm not deluding myself about seeing solutions or change; I can't afford to be that naive at this stage. Yet as a human I needed some energy pills to be able to continue. At one point I wanted to give up everything, I was so disillusioned. I just felt that the dialogues were going nowhere. They were circling around themselves, stuck. You kept on hearing yourself again and again, not breaking through. Even on the personal level, groups of women were talking but were not listening to each other; there was no substance to their discussions. We, the Palestinians, had no strength to offer to each other; we felt trapped and robbed by all these discussions and dialogues.

CHAZAN: Two things that are on totally different levels. First, I think that one problem with the discussions and dialogues in the past was really a conflict concerning Israeli women which we haven't articulated to ourselves. Our desire to be loved by you (and, by the way, Israelis have a dreadful need to be loved) made it difficult for you to separate the fact that we, the Israeli Jews, are part of the problem. And therefore I always sense justified disappointment in the fact that we couldn't do more about what we were complaining about than sit and talk to you. As you just said about the elections, I think that now we are saying, 'Ah ha, finally maybe I've done something to change the situation.' That's very important, because saying that changes the climate, and when the climate is changed maybe other things will become possible. This is deeply felt wherever you go: I see people standing up straighter; and there is a sort of lingering smile on people's faces.

MAR'I: You should see the reactions of my neighbors to the news. I'm now talking of the Jewish neighbors: everyone looks at me and says, 'We won.'

CHAZAN: But having said all of that (and let me note very quickly for the record, because I've been very much involved in this change, that it is ten days after the election and so we don't yet have any government, and I know it's going to be a long haul), I must also say that I've sat back over the last few days and thought to myself: What has created this change in mood? The difference between this feeling and something that would have been totally different, something which would have put us all, I think, into the deepest depression imaginable, is just two seats in the Knesset. Even if it's some kind of fluke, I'm perfectly happy for a change to come on the right side (or is it the left side?). But in reality not very much has changed – and I just hope to

God that we have the wisdom to nurture the mood so that something substantial will now happen. The sense of responsibility is amazing because it is much easier to complain about what somebody else is doing, but when you have to do it yourself, if you make mistakes – and we are going to make so many mistakes – wow!

MAR'I: I heard second hand that Shulamit Aloni said, Now that we are going to be part of the government we do not have the luxury of not compromising. She is now faced with the reality that she, as a head of the party, will have to compromise if she wants to be in the government.

CHAZAN: We're into a whole new venture, into the art of compromising without giving up too much. It's going to be an interesting time.

MAYER: Since we don't have much time left for discussion, I'd like to shift our focus here: Could you both consider how issues of national security have evolved over the last twenty-five years, and how such issues have affected Jewish and Israeli Palestinian women?

CHAZAN: I think that in Israel there is almost a fixation or obsession with security, and that 'security' has become a buzz word that justifies and explains and guides whatever the government decides to do.

MAYER: How has that affected women?

CHAZAN: I think that as soon as you put security at the top of your agenda, either voluntarily or as a result of the situation, women are virtually always going to be adversely affected. Because 'security' means expanding your military strength, and as long as the army and the military remain more important than anything else, women and women's issues are going to be marginalized. In a very real sense Israel needs peace so much because we have to rearrange our priorities, and we can't rearrange our priorities as long as security is number one on the national agenda as it has been. It infiltrates civil society.

MAR'I: We Palestinians do not believe that security is a real issue. We believe that it is an excuse. We can't even grasp how under modern conditions, given Israel's military superiority over the entire Middle East (not just over her neighbors), the issue of national security can continue to be insisted on. It really seems to be a psychological issue, rooted in the psyche of the Jews.

CHAZAN: (You're generalizing us again . . .)

MAR'I: The issue of Israel's national security does not affect Palestinian women directly. Yet the whole situation is, of course, the result of that: because of the Occupation, Israel's sensitivity to all matters which have to do with the Palestinian minority (including our actions and our movements) has been affected. But until recently Israeli Palestinian women were treated less seriously in security matters than Israeli Palestinian men. The authorities didn't really look at us as a potential threat to national security. I remember a women's group that we started in 1975: at first the authorities really thought it was great, and they encouraged us; but then all of a sudden they started hassling us verbally. One of the first warnings I got was from the municipality: 'Don't think that just because you are a group of women we believe that you are naive. We know what you are doing, we know what you want to do.' I asked, 'What do we want to do?' And they answered, 'Start an Arab university.' I responded that if an Arab university would actually endanger Israel's security then I wish we did have that dream and that potential and the ability to do things that you think are dangerous.

MAYER: Then security does create an interesting dynamic. There is a need to control 'them' because 'they' can constitute a 'fifth column'.

MAR'I: That is true for the society as whole. And that is why it is harder for me to differentiate women as a separate group. While as part of the whole society every minute of our lives has definitely been affected by security, security has not directly affected our status or women's issues. However, we have had a little more freedom of movement than our men have had, because the issue of security has made Israeli authorities try to control Palestinian men more and fear Palestinian men more than they have feared Palestinian women. And that is a big difference.

One damned blessing is that the Israeli authorities are still treating Palestinian society with the mentality of fifty years ago. It's amazing, as if they don't want to see the changes happening on all levels. They still don't look women in the eye, because they don't want to deal with another phenomenon. They just deny our existence – which gives us opportunities to act.

Again I don't want to sound as if I'm looking at the Occupation and at security issues from only the positive side, but I also have to be honest about the advantage of not being taken seriously by the Israeli authorities. It is still much easier for them to look at us as subordinates of our men. When, for example, I was being questioned at the Department of Arab Education, the interrogator confronted me and said, 'I'm really surprised and amazed. In the past, when we used to tell a Palestinian father about the misbehavior of his daughter, he used to slap her in the face, and she would just close her eyes without even moving. Where are those days?' I said, 'They're gone. They have been gone for a long, long time. Wake up, man!' That happened just ten years ago. This man probably thought that a Moslem woman who was called to Jerusalem by one of the deputies at the Department of Arab Education would be shaking with worry. He thought that I would be apologetic and would ask for forgiveness whether I had actually misbehaved or not. When I did not act that way he was absolutely confused. He couldn't accept this new reality, and he kept circling around it. I think that really typifies the way the Israeli authorities have been dealing with Palestinian women. And this, paradoxically, gives us some sense of freedom to act.

MAYER: What do you think about the joint efforts of Jewish and Palestinian women within the Green Line to end the Occupation, like Women in Black or *Dai LaKibush* ('End the Occupation')?

CHAZAN: There have not been enough of these efforts, in my opinion. (But it is very difficult to testify about yourself. I hope we are not involved here in total self-analysis; at this juncture I think it is much too early.) I think there has been a fair amount of joint activity, and it has been heavily politicized on both sides. In a funny kind of way, the fact that Mariam and I are sitting here together this morning is a result of these activities. I don't think that we would have ever met if we hadn't been involved separately in them. You see friendships and you see animosities across the communal lines, and that's marvelous, I think, on both a political and a personal level. But I'm not sure that we've succeeded in developing a leadership yet.

MAYER: A mutual leadership, or a leadership for each of the groups?

CHAZAN: First of all, a mutual leadership is a difficult thing to define. I don't think that we have succeeded in developing a core leadership from the grassroots. From that point of view I will talk about myself, but I think that Mariam also fits this pattern quite well. I stood on street corners with Women in Black and even initiated some activities, but, partly because of my profession, it would not be accurate to say that I sprang from below. I think that this activity strengthened certain women like me, but I still do not see a leadership which sprang up from within, or I only partially see it. And the truth of the matter is that until women have been active at the grassroots level and have nurtured their own leadership (and that's part of our responsibility, though it may sound very patronizing), we won't really have achieved much. One movement that both Mariam and I have been involved in from the beginning is *Reshet*, and because I played a central role in initiating it I know that it was also very elitist.

MAYER: What is *Reshet*?

CHAZAN: The Israeli Women's Peace Net (*Reshet Hanashim Lekidum Hashalom*), a broad coalition of women committed to a two-state solution, was founded in 1989. The women who have been involved in *Reshet* account for six of the eleven women members of the Knesset. On one hand, this is just an unbelievable amount of success. On the other hand, this is precisely the problem that I've been speaking about, for I don't know how deeply that success has penetrated to the grassroots.

MAYER: What all this really means is that there is a political voice as a result of women's organized attempts to advance peace.

CHAZAN: Yes, there is a political voice – but there is also an obligation to do more to cement the connection with the grassroots. I see that as one of my challenges. In the last ten days I've been making a list of challenges, so right now I'm overwhelmed.

MAR'I: I see the issue of dialogue between Palestinians within the Green Line and Israelis a little differently. I think that this is something which has not happened. Or if it has happened, it has been at the 'salon' level of exchange (which we probably all want to get to – but not to start from, because our fundamental problem is not social). On the personal level, of course, you can like an Arab and an Arab can like you or dislike you. That is a different story. But what I'm trying to say here is that the issue is not social but rather political. The process which groups like *Gesher*[5] have gone through could probably illustrate my point: initially their meetings centered around socializing, but after careful evaluation members became more politicized and the focus of their meetings became more realistic as well. Now they deal with the *real* issues.

The real challenge for discussions between women's groups could have actually been taken up by *Reshet* or by a group within *Reshet*. Instead, it has been a *fisfus* [a 'misfired' effort]. The main objective of *Reshet* was actually to dialogue with Palestinians on the West Bank. But whenever there were meetings inside the Green Line, whether Palestinians from within the Green Line were there or not, Israelis focused on how to clarify things among Jews themselves. As a result, we felt out of it. Although I have personally felt like a part of *Reshet*, sometimes when Israeli Palestinians and Israeli Jews are meeting together I feel that I'm the bridge between them, and at other times I feel that I am, as we say in Arabic, *Maflouha*, or torn apart. This has been a very interesting dynamic and I'm not exactly complaining about it, since *Reshet* really has been dealing with essential issues of dialogue with the Palestinians in the Occupied Territories. I am critiquing *Reshet*, however, for not fully challenging the status quo, for overlooking issues concerning Israeli Palestinians, and for not preparing the leadership that Naomi has called for. I feel comfortable with this critique precisely because I have been part of *Reshet*. As a member myself, I can say that the relationship between Palestinian Israelis and Jewish Israelis has not been on the agenda of *Reshet*. It has simply been taken for granted, without any real examination, that we are partners in dealing with the larger conflict.

CHAZAN: You raise some very interesting questions, which make a good place to close. What's going to happen to the peace movement in this country now that there is a possibility that the peace process could become official? We're real experts now in opposition, protest, demonstration, and criticism – but what's going to happen to the organizations within which we have matured politically? This has been worrying me a lot, including in the specific context of *Reshet*, because I know that in a sense *Reshet* will have to redefine itself and I'm not going to be able to take part fully in the process. I agree with you that there has been so much emphasis on Palestinians in the

Occupied Territories (who did succeed in establishing a women's leadership) that we haven't done enough among ourselves. But precisely when the formal emphasis shifts to dealing with, or hopefully resolving, some of the key issues of the conflict, then maybe the peace movement should start focusing on Arab/Jewish relations within Israel. Maybe that will be our next objective, if we can be mildly successful in the next fifteen years. (laughter)

MAR'I: The time is right now. Although, as I've been saying, there has so far been either a denial or lack of focus on the issues, perhaps we have had so many *fisfusim* because the right moment has not yet arrived.

CHAZAN: I agree with you. I think it is a constructive direction to go in, and I would like to pick up in the future with some discussion about the political *fisfusim*.

EDITOR'S NOTES

1 Professor Naomi Chazan is Knesset member of the *Meretz* party. *Meretz* is a combination of three left-wing parties, *Ratz*, *Mapam*, and *Shinui*, which together formed the Peace Block for the 1992 Israeli elections. In order to form a coalition, Prime Minister Rabin called on *Meretz* to join forces with Labor and establish his government. In return, four members of the Peace Block party have been appointed as ministers in Rabin's government.
2 The organization for Akka women was established in the 1980s as an educational organization for Israeli Arab women, particularly of Akka.
3 Akka, also known as Acre and Akko, was an Arab town and since 1948 has been a mixed Jewish Palestinian town.
4 Yossi Sarid, a Knesset member from the Citizens' Rights Party (*Ratz*) who has advocated talks between Israelis and Palestinians and who has long called for a two-state solution to the Palestinian–Israeli conflict, expressed his dismay with the Palestinians' support of Saddam Hussein's invasion of Kuwait in an angry chapter in one of Israel's leading newspapers (*HaAretz*, August 17, 1990 and again on January 29, 1991). In this chapter he stated that after supporting Saddam Hussein the Palestinians could get lost. Sarid's chapter initiated a fierce debate in the Israeli media about the relationship between the Israeli Left and the Palestinians, and between Israelis in general and their Left.
5 *Gesher* means Bridge. It is a Jewish–Arab women's group founded in the 1970s in the Galilee, incorporating women from all over the country. The group's aim has been working towards co-existence.

3

BETWEEN NATIONAL AND SOCIAL LIBERATION

The Palestinian women's movement in the Israeli occupied West Bank and Gaza Strip

Souad Dajani

By the early 1990s, the women's movement in the Occupied Palestinian Territories found itself poised at a new threshold. The Women's Committees, established in the mid-1970s, had reaped the rewards of years of diligent efforts to recruit and mobilize women. During the *intifada* especially, many of these efforts paid off as women assumed active, public roles in the struggle against the Israeli Occupation. At the same time, women's continuing activities in their communities resulted in a backlash in certain circles and raised new questions about women's roles, highlighting dilemmas that have confounded national liberation movements in many parts of the world. For years, as this chapter will indicate, women have been active in the Palestinian national struggle and have, for the most part, tried to locate their own agenda within the nationalist struggle as a whole. Yet now women are also coming to perceive the need for a specific women's agenda, distinct from the national movement.

Much of the following analysis concerns the period of the *intifada*. This uprising is characterized, at once, by the increased base of women's participation in the struggle against occupation and by the survival of traditional social obstacles to further political involvement by women.

This chapter begins by tracing the emergence of an organized Palestinian women's movement in the Israeli-occupied West Bank and Gaza Strip. It briefly touches upon the conditions under occupation that led to such an initiative, and evaluates women's efforts in both national and social terms. Rather than examining these issues through a progression of historical events, the chapter focuses on how Palestinian women themselves have perceived their roles, and on the directions they are taking in theorizing about the issues of social and national liberation.[1]

It has become almost axiomatic among national liberation movements that the struggle to end foreign rule takes precedence over other social agendas, including those of women. Palestinian women have largely shared this view. They have defined their activities and goals largely within these parameters, so that their efforts to improve their public roles and to expand their opportunities in society would be defined and legitimized in light of their contribution to the overall national struggle.

This has not meant that they have been unmindful of their specific concerns as women. On the contrary, Palestinian women have always evinced a specific women's consciousness and have sought to formulate strategies that would address concerns specific to women. Over the long decades of the Palestinian struggle, both before and after the 1967 Occupation, Palestinian women have redefined their agendas both in light of the changing circumstances that have impinged upon women's lives and in light of the general direction of the national resistance struggle itself.

Today, in the early 1990s, Palestinian women are becoming acutely aware of their vulnerability. They realize that national liberation is not necessarily synonymous with social liberation. They want their political activism, during the *intifada* and beyond, to be translated into real social gains and democratization throughout the whole of Palestinian society, so that they do not find themselves, as women, relegated to a permanently subordinate position after national liberation. Palestinian women frequently cite the example of Algeria (Kuttab 1992), where women participated extensively alongside men in the national liberation struggle only to find all their gains washed away after independence. Palestinian women are adamant about not allowing the same fate to befall them. Other social and national liberation movements have faced similar dilemmas and have raised similar questions about the links between national and social liberation; about how women should theorize about these issues; and about how the women's movement should move from theory to strategy in implementing its priorities. What is perhaps unique in the Palestinian case is that women there can now take advantage of the lessons and legacies of past struggles, to take informed steps in shaping the future which they envision.

This discussion of the national liberation movement is confined basically to the Occupied Palestinian Territories.[2] The Palestine Liberation Organization (PLO), which symbolizes the total struggle of the Palestinians, is located outside these areas. Within the PLO, women are represented officially through the General Union of Palestinian Women (GUPW). The PLO's component factions are prominently reflected in the Occupied Territories, where different grassroots and popular organizations claim allegiance to one or another of the four main political factions: Fatah; the Popular Front for the Liberation of Palestine (PFLP); the Democratic Front for the Liberation of Palestine (DFLP); or the People's Party (formerly the Communist Party). The various women's committees also reflect these factional divisions – and in a situation where the PLO cannot operate officially, it is the indigenous organizations and popular committees (of women and others) which take the lead in planning for resistance to occupation. Finally, both within and outside the Occupied Territories, changes in the means of resistance have reflected the modification of political goals, especially the official acceptance by the PLO of a two-state solution to the Palestinian/Israeli conflict. Over the two last decades, the PLO has moved from an almost total reliance on 'armed struggle' to engagement in political and diplomatic initiatives as legitimate forms of resistance in their own right. It was on the basis of similar assumptions that the *intifada*, the popular and largely nonviolent civilian uprising against Israeli Occupation that has been going on since December 1987, was launched.

THE HISTORICAL ORIGINS OF THE PALESTINIAN WOMEN'S MOVEMENT

A number of comprehensive analyses of the origins and early history of the Palestinian women's movement are already available (Abdo 1987; Giacaman n.d.; Jammal 1985; al-Khalili 1977; and Mogannam 1936). There is, therefore, no need to review these events here. It remains important to point out, however, that since its inception, the Palestinian women's movement has always been closely tied to the national issue. Activism on the part of Palestinian women occurred early during the period of the British Mandate in Palestine, in direct response to the threat posed by Zionist colonization. Women formed delegations to intercede with the British authorities; held congresses and other public events; and engaged in various forms of protest against the influx of Zionist Jews to Palestine and against the British policies that encouraged Jewish immigration.

Zionist immigration increased the dispossession and displacement of indigenous Palestinians, culminating in their forced exodus following the establishment of the State of Israel in 1948. Even earlier, the presence of a new sector of immigrants committed to the specific agenda of establishing a state for Jews led to an inevitable challenge to, and subsequent collapse of, indigenous Palestinian social structures. It was this condition – the inability of traditional Palestinian clan and family structures to sustain themselves as viable social, economic and political entities in the face of this forced disintegration – that led to the 'freeing' of individual members from such traditional ties. Since women were also affected by these changes, their awareness of their social and political conditions was consequently enhanced. They too were pushed into the public sphere and forced to resist the destruction of their world (Dajani 1993).

These events dramatize both the continuing integral link between the women's movement and the national movement and the manner in which structural changes interact with ideologies, beliefs, and world views at different points in time. Both of these issues have important implications for the present state of the Palestinian women's movement in the Occupied Territories: the first underscores the difficulties and contradictions that women face in trying at once to define a separate woman's agenda and to connect that agenda to the Palestinian national liberation movement; the second serves as a response to those within or outside the women's movement in the Occupied Territories who sometimes express frustration with the pace of change there. It is clear that just as an awareness of the national issue and of the role women would play within the national movement had to develop within the context of a structural challenge to Palestinian society, so too did a heightened awareness of the specific issues concerning women depend on broader social transitions. Indeed, the birth of what may be termed a feminist consciousness and agenda had to await the arrival (or imposition) of a new set of structural challenges to Palestinian society. These challenges came with the *intifada*.[3] We can, therefore, see a direct connection between the period before 1948 and that which unfolded after 1967. Throughout these years Palestinian women have been involved in a constant process of reacting to and acting upon their environment, a process which has culminated in the situation in which they find themselves today.

PALESTINIAN WOMEN AND THE NATIONAL LIBERATION MOVEMENT

It was largely in response to the political threat to their society that Palestinian women first mobilized to assume active public roles. Educated urban middle-class women played a significant role in developing the kind of social welfare activities that came to dominate a large part of the history of their movement, especially in the aftermath of the disintegration of Palestine. For example, in 1921 the first Arab Palestinian Women's Union was set up in Jerusalem, and in subsequent years various committees and groups were created to respond to emerging social and national concerns. Meanwhile, rural women were more directly active. There is evidence that as early as 1884, even before any organized women's movement had emerged in Palestine, women in rural areas were already struggling alongside their menfolk to resist the first Zionist settlements (Jammal 1985).[4]

The concept of 'relief', and the distinction between urban and rural areas, prevailed into the early period of Israel's occupation of the West Bank and Gaza Strip and the advent of the women's grassroots committees in the mid-1970s. The social activism of urban women that first emerged in early Palestine continued to be central during the post 1948 and 1967 periods. Thus, once the Zionist enterprise had culminated in Israeli statehood and Palestinians had been expelled and dispossessed altogether from their homeland, middle-class Palestinian women assumed chiefly 'charitable' roles. They were active in organizing the kinds of services and support networks that would be needed by the hundreds of thousands of landless and poor refugees scattered in exile.[5] At the same time, these efforts acted as a barrier to more extensive mobilization among women, either as active agents of resistance against Israel or as agents on behalf of a women's agenda. During this earlier period it was the national issue – the threat to Palestinian lands and livelihood, and to the very identity of Palestinians – that took precedence. This threat was immediate and tangible, and it had to be resisted at all costs. An awareness of a specific 'feminist' consciousness in the Western sense was clearly absent. When Palestinian women challenged patriarchal structures, they did so in order to fashion a political response to their common enemy outside, rather than directing their challenge internally, into their own society. One example of women's activism on behalf of the national cause was the first Arab Women's Congress of Palestine that was held in 1929. This event was described at the time as a 'bold step to take', in view of traditional restrictions against women, yet it did not directly articulate challenges to such restrictions (Antonius 1981: 8).

It was perhaps the continuing process of Israeli colonization that most dramatically transformed heightened political consciousness among women (as among Palestinians generally) into active resistance against Israeli rule. Following the 1967 June War, the remaining territories of what had formerly been Palestine were captured and occupied by Israel. Once under Israeli control, Palestinians were faced with what they perceived as a policy of 'creeping annexation' (Israel's official policy of structural integration without official annexation). Several processes underlay this policy, all of which contributed significantly to the destruction or transformation of indigenous Palestinian social structures and social relations in these areas. Israel moved immediately after the war to annex East Jerusalem and, in 1981, passed a 'Basic Law' that declared Jerusalem a unified city and Israel's eternal capital, effectively separating East Jerusalem from the rest of the West Bank and rendering it subject to Israeli law. (In contrast, until 1981, when the civil administration was imposed, the West Bank, like the Gaza Strip, was controlled directly by military law.) In the quarter of a century that has passed since 1967, some twenty Arab villages surrounding Jerusalem have had their lands confiscated to make way for the expansion of Israeli settlements around the city and the creation of a 'Greater Jerusalem' (PHRIC 1992). In the remainder of the West Bank and Gaza Strip, Palestinians have seen vast proportions of their lands confiscated, their trees uprooted, and their traditional livelihoods disrupted.

Although these effects of Israeli rule were not uniform within both the West Bank and Gaza Strip, Palestinians increasingly felt the consequences of economic and political strangulation: proletarianization; pauperization; and, most critically, the threat of their permanent uprooting from their lands and the disappearance of their Palestinian identity (Abu Ayyash 1981; and Graham-Brown 1984).

The Gaza Strip was the most seriously affected. Its indigenous social structures had already experienced disintegration in the aftermath of Israel's creation in 1948, due to a combination of factors: the influx of huge numbers of dispossessed Palestinians fleeing or expelled from Palestine; the slicing away of a major part of the territory of the Gaza Strip for incorporation into Israel; separation from the region's main trade sources; and, especially, the fact that the Gaza Strip had never possessed the indigenous economic structures (specifically in agriculture) that were necessary to sustain its population. Because its indigenous structures remained largely intact under Israeli rule and could continue to provide a livelihood for much of the population, the West Bank fared slightly better than the Gaza Strip. It was not long, however, before Israel instituted policies of land expropriation and settlement, restrictions against indigenous economic activities, and other measures designed to curtail and regulate virtually every aspect of Palestinian life.

One consequence of Israel's occupation policies was the proletarianization of Palestinians, as increasing numbers of both men and women were uprooted from their lands and livelihood and pressured to join the Israeli labor force in order to support their families. Many worked as poorly paid manual laborers, or as seasonal agricultural workers in Israeli enterprises.[6] Women from all sectors of society began to enter the education system, including higher education, in order to improve their chances of finding gainful employment. Women often took up jobs in the service sector, in teaching, nursing, and the like, or else received training in vocational skills which they could use to supplement their families' incomes. These socio-economic transformations inevitably acted back upon the women themselves. Increasingly, Palestinian women began to take upon themselves the task of defining their roles as productive members of Palestinian society under occupation and of developing avenues through which they would contribute to the struggle against Israeli rule. Given these conditions and what became a virtual daily onslaught against their Palestinian identity, it is not surprising that men and women alike regarded their collective national oppression as the first priority. Yet because Palestinian women's traditional structural position was different from that of men, gendered differences in Palestinian responses to Israeli Occupation did eventually emerge.

This politicization and mobilization occurred initially within the bounds of the sixty or so charitable societies that had been established within the West Bank before 1967. Under the Occupation, these societies expanded their purview from traditional welfare functions, to place greater emphasis on education, health and vocational training (Tunis Symposium 1984; Giacaman n.d). Perhaps the best known society of this type is *In'ash el-Usra* (Family Rejuvenation Society), which was established in El-Bireh in the West Bank in 1965. Headed by Samiha El-Khalil (the mother of Khalil, or 'Umm Khalil', as she is known) – an energetic, determined and nationalist Palestinian woman – this society exemplifies the older type of organization that has tried to adapt itself to changing circumstances.[7] An emphasis on self-help, as opposed to 'relief', was, however, largely absent in the more traditional charitable societies' framework. In addition, they were run mainly by middle-class Palestinian women and were located in the major towns and villages. These societies were, therefore, somewhat inaccessible to rural areas, and remote from these areas' growing needs.

In the mid-1970s, a new type of women's society began emerging to fill this vacuum and to address the changing needs of the population. By then, the imperatives of life under occupation had necessitated the questioning of traditional assumptions about women's roles and the expansion of the traditional definition of 'acceptable' women's behavior. Women themselves, newly confident, and radicalized by their growing involvement and activism in the public sphere, began taking charge and fashioning their own responses – both to their situation as women and to the Occupation itself. Just as the national movement in the Occupied Territories was moving away from passive *sumud* (steadfastness) to active resistance to Israeli rule, so too in the women's sector did the trend towards more active and more aggressive participation develop.

This is not to say that there had been no specific articulation of women's concerns in the intervening years. After all, the years since 1965 had witnessed the creation of the PLO and the establishment within it of the General Union of Palestinian Women (GUPW) – a direct acknowledgment of the specific contribution of women to the struggle. But in the 1970s, and perhaps for the first time within the Occupied Palestinian Territories, there emerged a distinct commitment to articulating women's concerns both separately and in connection with the national movement (though not necessarily in feminist terms at the time), and to working towards the establishment of an organized women's movement. Among the major concerns of these new organizations were: working women; women in rural areas; and the need to establish viable productive ventures and self-help opportunities for women.

The women who took the initiative to establish these committees were relatively young, educated, and activists in their own right. They believed that by strengthening the role of women in the national struggle they would be able to realize women's full potential in society. Their approach was to address daily concerns in various areas of women's lives, rather than to focus primarily or intentionally on political issues. They paid careful attention to involving local women themselves (for example, in rural areas) in expressing their own needs and priorities, and invited local women to participate in setting up and running branches of the committees in their own communities. The first of these grassroots women's committees was formed in the West Bank in 1978. Within just a few years, three additional committees had been established; each reflecting an affiliation with a particular faction within the PLO. The first to be formed, the Union of Women's Work Committees, was closely aligned with the Democratic Front for the Liberation of Palestine (DFLP). The Union of Working Women's Committees (1980) was associated with the Communist Party; the Union of Palestinian

Women's Committees (1981) with the Popular Front for the Liberation of Palestine (PFLP); and the Union of Women's Committees for Social Work (1981) with Fatah. These committees were distinguished from the earlier charitable societies mainly by their nontraditional, unofficial and grassroots character.[8]

Progress in involving local women was slow in many instances. The threat of Israeli crackdowns was ever present; and the unofficial nature of the women's committees made it difficult to rent spaces for their work, to staff and equip facilities, and to implement programs. Funding was a constant problem, as was the issue of competition and overlap among branches of committees with different ideological leanings. For example, it often happened that two women's committees of different ideological persuasions would compete with each other to set up a day-care center in a given locality. Another major obstacle to women's activism stemmed from Palestinian society itself and traditional views on the participation of women in public life. This was particularly true in the villages and rural areas (and some refugee camps), where women tended to be less educated and more enclosed within the confines of their communities. Activists of the women's committees had some difficulty in finding appropriate ways to communicate their goals to these women and to convince them that they, as well as their families and communities, would benefit from participating. This method of approaching political issues indirectly, by way of attention to daily concerns, gradually paid off. Women could convince their male partners that the activities planned by the committees fell within the range of 'acceptable' behaviors for women, and also received training in skills that would enable them to supplement family incomes or provide other services, such as in health, day care, and the like.

In expanding their activities in towns and cities Women's Committees were also beset by a number of concerns that were more specific to location. For example, urban women tended on the whole to be more educated and, therefore, more likely to be enrolled in the work force or more desirous of joining it. In these settings, the issue was how to ensure that women were protected from discrimination and inequality in the workplace because of their gender. In rural areas, the question was how to convince fathers, brothers and husbands to allow women to train themselves in order to work outside the home. In order to address these concerns and mobilize women more effectively, several of the Women's Committees focused specifically on the issue of working women. For example, the Women's Work Committees placed special emphasis on the rights of women in the work force; on their rights to leaves, vacations, better wages and, especially, representation in labor unions (Khreisheh in Najjar 1992; Hiltermann 1990; Siniora 1989; Women's Work Committees 1983; Tunis Symposium 1984). The issue of rights has always been problematic in the Palestinian context and has presented serious dilemmas for women activists. Most Palestinian unions have been dominated by men, and have also been subject to constant harassment by the Israeli authorities. In this general atmosphere of oppression and control, Palestinian men could not, or would not, appreciate the immediacy or relevance of women's concerns. Women themselves were often reluctant to demand their rights, partly because they may have feared for their jobs, but also because they did not want to alienate their male compatriots with whom they were engaged in a joint national struggle. Until recently, gender and class issues have generally assumed a lower priority in the national struggle, and they remain a source of debate within the Palestinian women's movement. It has only been in the years since the *intifada* that Palestinian women have begun to insist on the inclusion of a social agenda for women's liberation as a priority in its own right.

The role of Palestinian women in the *intifada*

It is difficult to appreciate the urgency with which Palestinian women approach the national and social issues that confront their society under occupation without examining women's roles in the *intifada*. It was this experience that institutionalized women's involvement in the national struggle; and it was the reaction against their activism that helped to crystallize women's social agendas. Between the time of their inception and the beginning of the *intifada* in December 1987, Women's Committees in the occupied West Bank and Gaza Strip continued to grow. This growth paralleled the expansion in the Occupied Territories of voluntary and grassroots committees as a whole, in response to the growing needs of the population under occupation for alternative services and support networks. The extensive involvement of women in these activities established the basis for the later participation of women during the uprising. The late 1970s and early 1980s were also a period of harsh repression under Israeli rule, during which the work of local municipalities became increasingly circumscribed and indigenous organizations faced multiple restrictions on their operations. Much of the impetus for the creation of voluntary and popular committees came specifically as a response to the election of the first Likud government in Israel, in 1977, and to its policies towards the Occupied Territories. Under Likud, land expropriation and settlement building increased in both pace and scope, and galvanized Palestinians into action to protect their society. Palestinians' response to the limited self-rule under continued Israeli occupation slated for them by the 1978 Camp David Accords between Israel and Egypt (brokered by the US) was an accelerated drive to maintain their *sumud* and to enhance it with deliberate efforts to develop their society. Israel continued to place obstacles in the way of any form of indigenous development. In 1981, after the civil administration was created, the Israeli authorities dissolved Palestinian municipalities and removed several municipal officials from their posts. Activists in other organizations were also arrested, deported, or otherwise prevented from carrying out their duties. Palestinian organizations – such as the National Guidance Committee (NGC), which had been formed to organize resistance to the provisions of the Camp David Accords – were officially banned. The popular committees that sprouted and grew in this atmosphere provided a legitimate and acceptable institutional framework for the increased involvement of women.[9]

This was a critical decade in terms of women's mobilization and participation in public life. As Israeli repression intensified, Palestinian women became acutely aware of their precarious and conflicted position under occupation. They were forced to assume new economic responsibilities in providing for their families, sometimes even as sole providers in the absence of male relatives who had been imprisoned, deported or killed by the Israeli authorities. These roles were added to their traditional responsibilities as wives and mothers, as the preservers and transmitters of the culture and identity of the Palestinian people. As the notion of 'motherhood' and sacrifice became enshrined as a virtual national duty, women who were struggling for recognition in the public arena were forced to somehow combine this traditional view of their duties with their newly defined roles of providing economically for their families.

Much has been written about the *intifada* in general and, more particularly, about the role of women in it (Strum 1992; Abdo 1987, 1991; Abdul Jawwad 1990; Dajani 1990; Jad 1990; Taraki 1989, 1990; Warnock, 1990; Giacaman and Johnson 1989). The term itself comes from an Arabic word that signifies a 'shaking up' or a 'shaking off'. Indeed the *intifada* has lived up to its meaning, involving both the shaking off of the Israeli Occupation and the shaking up of Palestinian society itself under Israeli rule.[10] Intentionally or not, the second element has become dominant over the course of the uprising.

The distinctiveness of the *intifada* has derived from its popular appeal across the Occupied Territories. Within just the few initial weeks of the uprising, the mass-based grassroots committees had organized a web of local administrative committees. These committees proliferated throughout almost all of the Occupied Areas. The idea that Palestinians could be so effective in mobilizing themselves in their communities was virtually unprecedented, and was a significant factor in Israel's initially confused response. Numerous Palestinian uprisings had occurred in the past, but never on such a scale or with such a degree of popular determination and solidarity. The *intifada* indicated that the people under occupation were taking matters into their own hands (with the evident approval of the PLO, outside), and challenging Israeli rule. In other words, the locus of power had shifted away from the PLO to the people themselves who lived under Israeli military rule. Simultaneously, civilian resistance emerged as a separate and legitimate means of struggle. This, too, was a radical departure from the earlier periods, when only 'armed struggle' and, generally, some degree of 'political and diplomatic' efforts were recognized as legitimate forms of resistance by the Palestinian national liberation movement. As became clear in the course of the *intifada*, this emphasis on resisting the Israeli Occupation by means of a civilian resistance struggle had both positive and negative consequences for all concerned.[11]

Within the Palestinian community under occupation, the *intifada* did indeed succeed in 'shaking up' traditional structures and established social relationships. People quickly mobilized and directed their efforts towards the national goal. Not only in the grassroots committees but also within the different institutional sectors, Palestinians attempted to adapt their traditional roles to meet the new challenges of the national struggle. Through such action, Palestinians were sending a signal to Israel that they would refuse to cooperate with the Occupation regime and that they would attempt to establish their nongovernability by Israel. They hoped that direct action at key levels would place increased pressure on Israel to reevaluate its continued rule over Palestinians, and eventually force Israel to withdraw from the Occupied Territories. Some of the existing indigenous economic enterprises, for example, turned their production to the support of the *intifada* by producing goods that had previously been unavailable in the Palestinian community and therefore had to be purchased from Israel. This tactic helped to uphold the boycott of Israeli products that had been called for early in the *intifada*. At the popular level, 'home economies' became the non-institutional response, as Palestinians claimed empty plots of land in their communities on which to grow vegetables or raise chickens and rabbits. When all the universities in the Occupied Territories were closed by military decree in February 1988, Palestinian educators managed to set up 'popular education committees' which met with small groups of students in homes, mosques and churches across the occupied areas.[12]

Palestinian women played a pivotal role in the *intifada*. Building on the experience they had gained during the previous decade in the women's committees, Palestinian women were able to extend their activism and newly found organizational skills in the service of the uprising; to help mobilize people in different communities to perform strategically important functions, such as setting up units to collect and store food and establishing guard units to warn their community in the event of the approach of Israeli soldiers or settlers. Women helped to establish alternatives to the organizations of the Occupation regime, a goal that was central to Palestinian efforts to 'unlink' from their dependence on Israel: they were able to team up with the Medical Relief Committees to provide needed health care; and they worked with the Agricultural Relief Committees to reclaim small plots of land for the 'home economies'. Palestinian women also participated more overtly and directly in the *intifada*. They held their own demonstrations, threw stones alongside youth, and engaged in other protest actions to express their rejection of Israeli rule. They exhibited tremendous courage in their confrontations with Israeli soldiers, frequently risking their own lives to rescue *shebab* (Palestinian youth) and children who had been caught by Israeli soldiers or police.

It was this particular feature of the *intifada* – the widespread participation of women in the struggle against Israeli rule – that, perhaps more than anything else, dramatizes the extent to which Palestinian society had itself been 'shaken up' by this struggle.[13] For the *intifada* proved itself to be a double-edged sword, both for the Palestinian community in general and for women in particular, who defied traditional restrictions to participate in the popular resistance efforts. There is a clear connection between women's increased activism and the general atmosphere of social change generated by the *intifada*. At one level, this can be attributed to the failure of indigenous social structures, including the family, to perform their traditional functions in Palestinian society. Although this process had begun long before the uprising – due to the impact of the Israeli Occupation on Palestinian society generally – it was exacerbated by the uprising. In addition, the courage and activism of sons and young men (the Palestinian *shebab*) leading protests and demonstrations against the Israeli Occupation and the respect they earned in their communities – at least in the early stages – served as an example to women. Soon, daughters, sisters and wives were also challenging patriarchal restrictions to their public participation in the uprising. It was clear that the *intifada* was a national effort – a kind of higher calling – that required commitment and sacrifice, and removal (or suspension) of traditional customs and values. There was a spirit of rejuvenation throughout Palestinian society: people began talking and planning, not only for national independence, but for the type of society in which they would live in the future. Many in the West Bank and the Gaza Strip talked about the 'democratization' of Palestinian society, and hoped to combine strategies for national liberation with strategies for social liberation. At the forefront of the struggle for equality, democracy and freedom were many Palestinian women.

These processes of reevaluation and of theorizing about the future of the *intifada* and the forthcoming Palestinian state did not, however, occur at once. Even the revolutionary situation does not necessarily or automatically advance gender equality. Largely responsible for fashioning their own new roles, Palestinian women have nevertheless stopped short of issuing a direct challenge to the patriarchal structures. They have preferred instead to expand the boundaries of their involvement within these structures. In their view, the danger to the nation precedes their own priorities, and they feel the need to struggle alongside men against this external threat. Moreover, many feel that this is an inappropriate time to be alienating their male compatriots. In this, Palestinian women have been no different from other nations and their liberation movements. In Algeria and South Africa, for example, even as women assume more active public roles they are still entrusted with the responsibility of 'motherhood' (in much more than its biological sense) and the preservation of the nation's identity, and thus bear what is commonly referred to as the 'triple burden'. Still, there comes a time in all revolutionary movements when the processes of change cannot be contained and contradictory forces cannot be reconciled. Within the Occupied Territories, this point came with the emergence of the extremist Islamic groups, notably *Hamas* (The Islamic Resistance Movement) in the Gaza Strip, and the subsequent religious fundamentalist backlash against women that took place during the *intifada*.

The rise of the Islamic movement came at a time when women were already experiencing restrictions on their participation in the *intifada*. By then, around the second year of the uprising, Israel had already banned the popular committees, which had served as such a powerful forum for the induction of women into political activism. The demise of these committees left fewer opportunities for women to enter public life. At the same time, the *intifada* had begun to lose its earlier momentum. Palestinians under occupation perceived that they were making enormous sacrifices without achieving comparable political returns. Political factions had lost the unity and consensus of earlier months: they differed on the course of the uprising, on the means of struggle, and on the interpretation of Israeli and American moves to address the situation.[14] The killing of suspected collaborators was on the rise, and Israel was forever devising new ways to split and control the national movement and to break the will of the people to resist. In this atmosphere of despair, anger, and frustration with the political process, the message of the more radical wing of the Islamic movement resonated among Palestinians in search of meaning and hope (Hammami 1990; Legrain 1990; Abu Amr 1989; Taraki 1989). As in all national struggles, culture – including religion, language, tradition and custom – had a powerful appeal. Indigenous culture could be mobilized effectively as a response to the 'alien' culture of the opponent. Ironically, *Hamas* owed its rise primarily to the Israeli authorities, who had earlier courted the movement as an alternative to the PLO.[15] In some areas of the Occupied Territories, particularly in the Gaza Strip where conditions were especially harsh, the Islamic movement gained widespread support, sometimes shaped less by religious beliefs than by the practical considerations of survival: in a situation characterized by the breakdown of traditional social supports, *Hamas* was well positioned to provide support services and relief to the local population.

The message of *Hamas* and other radical Islamic groups was simple: Israel was winning because Palestinians had abandoned their culture and religion; in order to prevail, Palestinians would have to return to Islamic tradition and abandon Western practices and values. The essence of this attack was directed at women. *Hamas* and other groups urged women to forgo public activities and return to their homes, to resume their traditional roles as wives and mothers. They urged women to don the veil, to segregate themselves from men, and to allow men to conduct the struggle against Israel.

As the appeal of extremist religious groups spread from the Gaza Strip to areas of the West Bank it became an important source of concern for Palestinian women activists. Women activists realized that 'democratization' had to be more than a slogan that was appropriate only during the period of national resistance and would then be cast aside after the establishment of an independent state. They also realized that achieving national liberation without social liberation would be turning back the clock and giving up all that they, as Palestinians and as women, had gained over the long years of sacrifice and struggle.

For the most part, the nationalist leadership in the Occupied Territories disregarded women's growing concerns over the rise of Islamic fundamentalism. These Palestinian leaders seemed to think that the forces of fundamentalism were transient and could be contained within the national struggle, especially once political gains could be demonstrated and attributed to the secular leadership. Most of all, the nationalist leadership[16] tended to minimize its reactions so as not to unduly antagonize the religious movements. Ultimately, these leaders placed political considerations above social concerns, and decided to risk alienating women and marginalizing them within the struggle rather than to incur the wrath of the powerful religious movements and to open themselves up to accusations of being traitors to the national cause and enemies of tradition.

This policy of trying to steer clear of the issue of women and religion clearly backfired. Fundamentalist backlash against women became overt and severe; there were even cases of women in the Gaza Strip who were beaten or who had acid thrown at their legs for daring to walk in the streets unveiled. Pressure mounted on the nationalist leadership to address the issue of fundamentalism publicly and directly, and some male leaders, such as Faisal Husseini, did respond.[17] But this whole event shocked women irrevocably into an awareness of the discrimination and oppression which they could face in a future Palestinian state. From this point on, one can begin to speak of a genuine feminist consciousness emerging among women in the West Bank and the Gaza Strip.

FEMINISM AND GENDERED POLITICS IN THE OCCUPIED PALESTINIAN TERRITORIES

A single public event unmistakably marked a turning point in the development of a genuine feminist consciousness in the Occupied Territories: the conference on women that was held in the West Bank in December 1990. There, for the first time, women activists decided to broach the subject of religious fundamentalism publicly and critically. Despite potential dangers, some five hundred women from across the Occupied Territories attended. Prominent male leaders were present as well: Faisal Husseini, for example, gave a keynote speech. The appearance of men at this conference helped to give an official stamp of approval to the efforts of women and, more importantly, to indicate publicly that backlash among extremist religious groups against women was unacceptable and would have to be addressed immediately and directly at the highest levels of the national leadership. From this point on, Palestinian women began to devote increased attention to developing a specific women's agenda and to formulating and implementing a program of social liberation. While they realized that successfully steering a path between achieving women's goals and avoiding excessive antagonism of their male compatriots posed a thorny problem, 'We will not be another Algeria' remained a rallying cry among women activists in the Occupied Territories. In order to guarantee that the struggle for national liberation would incorporate full democratization of Palestinian society, Palestinian women began to theorize more carefully about the issues confronting them.

These efforts are still underway in the Occupied Territories. In the end, the real test of their successes will lie in their ability to translate theory into practical plans, and strategy into reality. Among the many issues that Palestinian women have addressed in this context, two stand out: the question of feminism; and theorizing about social agendas and national liberation.

The question of feminism

The meaning of feminism from the standpoint of Palestinian women, and the problem of incorporating it into a women's agenda for social liberation alongside national liberation, is an immediate and contentious issue in the Occupied Territories.

Many observers would agree that only in the last two years or so, since about 1990, has a feminist consciousness discernibly emerged in the West Bank and the Gaza Strip (Gluck 1992; Strum 1992). This development constitutes a major departure from the past, when Palestinian women were often loudly insistent that 'feminism' was an alien Western construct that was not applicable to their concerns.[18] More recently, however, Palestinian women have begun to refer openly to themselves as feminists and to debate feminist analysis. Although they are publicly broaching subjects that would have been unthinkable just a few years ago (including wife-beating and other types of abuse; abortion; and women's health movements), Palestinian feminists are not primarily concerned with the individual manifestations of oppression and discrimination but, instead, with the overarching societal framework that is embodied, principally, in the Personal Status (Family) laws.

In most of the Arab world there is no effective separation between religion and state. For centuries, even before these countries gained independence (as could also be seen in Palestine under Ottoman rule), each community ran its 'personal' affairs according to its own particular religious tenets covering, essentially, marriage, divorce and inheritance (Hijab 1988). Following the establishment of separate nation-states in the region, this system persisted and became institutionalized into law. Whereas civil law was made applicable to other state and community affairs, most of these countries administered 'personal' affairs on the basis of religious law, frequently Islamic law or modifications thereof. Palestinian women are now concerned that the future Palestinian state will simply emulate its neighbors and will not be democratic. On the matter of polygamy, for example, they maintain that since the declaration of the Palestinian state in 1988, Palestinian passports which have been issued in conformity with Jordan's, carry separate pages for men to list up to four wives.[19] In Islam, men are allowed to marry up to four wives under certain strict conditions, and in Jordan these matters are regulated by the Personal Status laws. Palestinian women fear that women will be legislated into an inferior status right from the beginning, on the basis of such laws, and they are determined to begin working now to prevent such an outcome.

Toward this end, Palestinian women have recently established several Women's Studies Centers in the Occupied Territories, including at least three in the West Bank, one of which has a branch in the Gaza Strip. One particular focus of the Bisan Center For Research and Development in Ramallah is the Personal Status laws (Strum 1992; Abdo 1990; Kuttab n. d.). Female researchers there are in the process of consulting with lawyers, religious figures and other professionals to develop a complete civil code that would govern personal affairs in the future Palestine.[20]

Theorizing about social agendas and national liberation

As this discussion of feminism clearly indicates, Palestinian women have begun to theorize about both connections and differences between social and political liberation. The real test of their efforts, however, will lie in the extent to which they succeed in articulating social agendas to be instituted in the Palestinian community before liberation. This would involve at least two main undertakings: Palestinian women must manage to develop a sound theoretical base for assessing their conditions; and they must be able to translate theory into real strategies for change. While this section will address only the first issue, the task of theorizing, we do need to keep in mind that theorizing remains, inevitably, only the first step towards the formulation of a coherent strategy.

Palestinian women have begun to make great strides in identifying the key variables in their social condition. Variations in interpretation have occurred over time as women have begun, for example, to assign more weight to gender issues in view of their own location in society, or to analyze from their own political, ideological and/or class positions. Other variables which have come into play include ethnicity – comprising both nationality and religion in this instance – and the state, primarily the Israeli state that continues to rule over them. Complicating matters further is the interplay among these diverse variables, and their intersections within Palestinian society and between the Occupied Territories and Israel. Palestinian women have made concerted efforts to sort out these factors in order to decide how they can be manipulated effectively in a program of social and national liberation.

One of the major dilemmas confronting Palestinian women in their struggle for democracy and freedom is the absence of a Palestinian state (Abdo 1991; Yuval-Davis and Anthias 1989; Joseph 1986). Though this void leaves a window of opportunity to initiate change, it also places women in the difficult position of having to chart a new course in defining their public and private roles and in deciding where their energies can be most effectively invested. Private roles, even traditional roles, become politicized, as in the emergent symbolism of the 'mother as martyr' and all the other images of a glorified Palestinian motherhood. Palestinian women cannot be blamed for stretching the definition of their traditional roles into these new politicized entities; indeed, their courageous and innovative responses reflect women's growing political consciousness and assertiveness. However, as women continue to emerge from the confines of existing social structures, these structures become subject to erosion both from within and without and the roles of women change, in turn, even further, as evidenced during the *intifada*. The absence of a Palestinian state also intensifies the significance of informal social or communal groups (as opposed to formal state structures) in the expansion of women's participation in social and political activities, as clearly demonstrated by the predominance of women in the local committees during the *intifada*. The importance of the informal sector in the Palestinian case is also indicated in the dramatic drop in the level of women's participation during the period after August 1988, when Israel issued a ban against the popular committees. For the informal women's committees had legitimated women's activism in the period before the *intifada*, and contributed to the expansion of women's participation during the uprising itself.

In the wake of the backlash against women during the *intifada*, Palestinian women currently are faced with a dilemma concerning their 'informal' participation: the absence of a Palestinian state at this juncture has perhaps made them more vulnerable, for there are no official structures in place to protect their rights. The lessons of the *intifada* indicate that it is not very likely that further mobilization of women in informal types of groupings will eventually translate into recognized and legitimized formal roles for them. Yet beginning work immediately on entering into more formalized public sectors and establishing themselves there officially, in order to ensure women's presence as equal to men's once the Palestinian state is established, also has its risks, not the least of which are the traditional obstacles to women's participation in the public sphere. Still Palestinian women have been attempting precisely that – to move directly into the public sector and thus to challenge traditional patriarchal structures and beliefs. Women have, for example, placed special emphasis on membership in trade unions, in order to gain their rights as women workers. As noted earlier, Palestinian women have also established a number of centers which are currently engaged in investigating the conditions of women's lives. And women have been involved in creating their own economic enterprises, in the form of women's cooperatives which they are totally responsible for administering and running; one prominent example is the Production is Our Pride food cooperative in Beitello (Strum 1992; Abdo 1990; Kuttab n.d.).

While it is important not to underestimate these efforts, it is also important to keep in perspective the theoretical concerns that they raise. Palestinian women seeking to get ahead in Palestinian society as it is presently structured may only be contributing to the reproduction of a similar class system in a newly formed state. So far, an emergent feminist consciousness seems to have triumphed, at least among a certain sector of women activists within whose analysis gender has taken precedence over class. Yet, implementing their analysis could result in a situation where only women of the elite classes would have advantages and a certain 'equality' in a Palestinian state. A real program of social liberation would require more concerted efforts to analyze and address class issues.[21]

In theorizing about their conditions and setting agendas for the future, Palestinian women must also address directly the element of religion, specifically Islam. Palestinian women have tried to assess how far they should go in challenging the religious underpinnings of their status in society. They realize that to struggle to become 'political' beings within the national liberation movement is often more acceptable than trying to challenge their 'private' roles within their families and within community structures. They question whether they should, as women, challenge patriarchal and religious structures directly, and thus risk alienating their male compatriots; or whether they should instead seek change indirectly, through continued participation in the public sphere in a manner that does not directly challenge traditional structures.[22] Some Palestinian women have urged that the political issues continue to take precedence, both in order to achieve ultimate liberation and in order to avoid splitting their movement and thus rendering themselves more vulnerable to Israel. Others, especially feminists, have come to the conclusion that they risk becoming 'another Algeria' if they do not work from the start to initiate change, no matter how upsetting and challenging this may be (Gluck 1992; Holt 1992; Strum 1992; Warnock 1990). Finally, Palestinians are locked in a national struggle against an intractable opponent. In theorizing about their conditions and about the possibility of combining a social agenda with a political agenda of liberation, Palestinian women have had to keep this reality firmly in sight.

While it is impossible to do justice here to the myriad factors that come into play in the relationship between Palestinians and Israelis, the ties that have developed between Israeli and Palestinian women over recent years are an important element to note. Israel too has its feminist activists. However, it is by no means a given that these two feminisms, Palestinian and Israeli, will converge. Many Israeli feminists are concerned solely with Israeli affairs, like equality within Israeli society, within the army, or in the realm of religious practices.[23] On the other hand, a significant sector of the feminist movement in Israel is concerned with the issue of peace, and is highly visible and active in this struggle (Hurwitz 1992; Sharoni 1992). To these women, peace with the Palestinians is inextricably linked with the struggle for equality and social justice for all women. Significantly, in recent years 'feminism' has constituted a common meeting ground or point of convergence for women of different ethnic backgrounds. In the Palestinian/Israeli conflict, the shared concerns of women struggling for social liberation have become a basis for establishing networks and solidarity groups which have, in turn, become a point of entry into the national question and motivated women to struggle jointly for a just peace. For both the Palestinian and the Israeli women who are involved in these efforts, social liberation will never be complete until the national conflict and Israel's occupation of the West bank and the Gaza Strip have ended, and increasingly, for both Israeli and Palestinian feminists, national liberation will never be complete without also incorporating programs which seek social liberation and justice for all (Abdo 1991).

There is an interesting process at work here. Because of their placement in society – often as victims of sexism, and racism – women are uniquely positioned to perceive the connections among different spheres of domination. As feminists they can then try to challenge the hegemonic versions of history that have kept them oppressed, invisible, and separated (Mohanty 1991). So begins the struggle to reclaim and legitimize – simultaneously – both the feminist and national liberation movements (Gilliam 1991).

There is no single answer to the dilemmas which confront Palestinian women. Palestinian women know that they want to be free of Israeli rule, but they also want to be free and equal citizens of their own state. Somehow, social and political liberation must go hand in hand: Palestinian women cannot afford to go back.

CONCLUSION

Where do Palestinian women now stand in their struggle? To answer, we must return briefly to an earlier question about moving from theory to practice; about, in essence, strategizing.

Some Palestinian observers, including feminists, have criticized the women's movement in the Occupied Territories for falling short of its goals.

Eileen Kuttab, for example, maintains that Palestinian women have not only failed to extend their public activism into real political gains or, generally, to translate theory into practice, but they have also not yet succeeded even in changing their traditional image in the Palestinian community (1992). Kuttab maintains that in spite of women's participation in the *intifada*, women have failed to generate real change.

Kuttab and others perceive a wide gap between theory and practice. This gap reflects in turn, the problematics of incorporating a struggle for social liberation within the struggle for national liberation. Philippa Strum goes further and claims that Palestinian women have not concentrated sufficiently on women's emancipation as a goal in itself, rather than as an element within the struggle for national independence (1992). In her view, Palestinian women have not theorized enough about the specific factors which they confront in their struggle for social liberation, both in conjunction with and separate from national liberation. Sensitive issues, such as the patriarchal structure and its ideological underpinnings – in which women have colluded, so to speak, in maintaining and reproducing their own oppression – have yet to be dealt with. Finally, in Strum's opinion, a thorough analysis of the connections between the national liberation movement and the women's liberation movement – and, most importantly, of how to achieve the latter before liberation, of how to strategize – has yet to emerge. As Palestinian women continue in their struggle, it may be useful for them to seek comparisons in the comparative successes of the present, like South Africa, rather than in the failures of the past, like Algeria. For all their concerns and warnings, there is in fact little danger that Palestinian women will succumb to the predicament of their Algerian sisters, since Palestinian women have already traversed far beyond the point originally reached by Algerian women in their struggle. There can hardly be a retreat into invisibility for Palestinian women now. The question is, then, how to go forward.[24] Despite important differences between them, Palestinians can benefit from black South Africans' long decades of struggle, especially by paying careful attention to the genesis and growth of the women's movement among black women in South Africa.

Like Palestinians, black South African women began to organize politically around the issue of 'national liberation'. In 1910, white settlers from four British colonies formed the Union of South Africa, an entity based explicitly on the principle of white supremacy.[25] Although the formal institutionalization of white rule under the Apartheid system did not take effect until 1948, the impact of the policies of 'white supremacy' was felt immediately by the native African population. From the start, the white settlers employed racial and racist ideologies to relegate black Africans to the status of a lower social class, as a cheap labor force for the emerging capitalist economy. The Apartheid regime formalized the racial and class separation between blacks and whites by restricting black Africans to various 'reserves', the 'homelands' from which black Africans commuted as a migrant labor force to work in the white-dominated South African businesses.

Initially the involvement of black African women in the struggle for liberation was limited, and was defined largely around the idea of 'motherhood', or 'motherism' as Cherryl Walker refers to it (1991). As in the Palestinian struggle against Israeli occupation, the concept of motherhood was neither static nor conservative, and black African women gradually began redefining their traditional and private roles into revolutionary and political ones. Though not 'feminist' at the outset, the reformulation of these traditional roles served to empower black women and to propel them into action and into an enhanced state of political and social awareness. The period of increased involvement coincided with economic changes under white rule and with the growing proletarianization of black African women alongside black African men.[26] Like Palestinian women, black South African women became preoccupied with the question of women and work, and with articulating their positions in terms of gender, race and class. Issues such as paid work, worker's rights and, indeed, the very right to work assumed increased importance in an atmosphere of dispossession and exploitation of the black population by a powerful opponent under the Apartheid regime. Black women in South Africa also had to grapple with the question of whether or not to join trade unions, and of whether their efforts to improve their position as women would antagonize their menfolk.

Ultimately, like Palestinian women, black women in South Africa faced the issue of theorizing about their conditions, and of moving from theory to strategy. They, too, wanted to avoid deflecting energy from the total struggle for an end to Apartheid. They wrestled with the dilemma of where to situate their movement; of whether the movement could or should be conducted within the main organizational framework of the national liberation movement, in this case the African National Congress. While the 'feminist' movement in South Africa has since its emergence been rent by factionalism and in-fighting among various groups over the goals and means of struggle (Walker 1991), over time women activists decided that they would focus together on power structures in South Africa located specifically within the black community, by organizing around issues that affected them directly as women. They would, however, insist at the same time upon organizing on a national scale.[27] As in the case of the Palestinian women's organizations, black South African women clearly recognized the roots of their oppression, both in the patriarchal structures of the black community and in the system of white rule – but they had trouble translating these theoretical concerns into a viable strategy. Ultimately, the black South African women's movement chose to focus on specific abuses suffered by women, and therefore to concentrate largely on creating services within their allotted 'homelands' and improving conditions there. Since within the Occupied Territories the question of whether to direct energies towards total national liberation or to concentrate, instead, on improving conditions and standards of living within the 'autonomous' Palestinian communities is becoming increasingly central both for the Palestinian community as a whole and for women in particular, the South African example may prove especially instructive.

Over the last few decades, Palestinian women have had to contend with a series of major upheavals in their lives, including the launching of the *intifada*, and the Gulf War and its aftermath. Women activists have not hesitated to seize upon the opportunities provided them, particularly during the *intifada*, to challenge further the definitions of acceptable roles for women. Their growing involvement in this uprising and the backlash that has emanated against them from traditional religious circles has only strengthened women's conviction that national liberation is insufficient without more fundamental social change, including equality for women and democratization throughout Palestinian society.

It was during the *intifada* that Palestinian women activists began to seriously investigate the possibilities for advancing a women's agenda alongside the struggle for national liberation. Palestinian women need to make the transition from theory to practice, and to translate their theoretical understandings into viable strategies for change. They have so far made remarkable strides in understanding the roots of their oppression and in making their own voices heard in their communities. They have succeeded in identifying the key variables – within traditional Palestinian society and under occupation – that define their social conditions. And Palestinian women have made significant advances in developing a theoretical construct that is capable of explaining these conditions and their implications for women in the future Palestinian state. Perhaps, indeed, as some have predicted, the next intifada will be the women's *intifada*.

NOTES

1 In this chapter, we will limit our discussion to Palestinian women who live in the Israeli occupied areas. There, Palestinian women are found in all walks of life: rich and poor; urban and rural; old and young; refugee and non-refugee. Some are known activists; others less so. Some face tremendous obstacles to social participation from within their own familial and social environments, while others enjoy the full support of family and community members. Palestinian women are sometimes the sole supporters of their families, in cases where their male relatives have been killed by the Israeli authorities or are absent due to imprisonment or expulsion. Women have themselves endured Israeli punishment: they have been imprisoned; and some have experienced beatings, torture, and other forms of mistreatment.

2 A note should be added here concerning the research methodology for this study. I find myself, as a Palestinian and a woman, sharing a common heritage with women of the Occupied Palestinian Territories. However, apart from two visits during the *intifada* (in January/February 1988 and October 1989), I have never

actually lived in these areas. My experience as an 'outsider', a city-dweller and a member of the middle class, inevitably colors my interpretation of events, and the degree of significance which I may accord any specific variable. However, being 'outside' may also have its advantages. It enables me to reflect on issues facing Palestinian women in a more abstract and dispassionate manner, and allows me to add my voice to the ongoing debates surrounding these issues.

3 Though this discussion focuses essentially on women, the same redefinitions are occurring in the wider political and social sphere, where we can see that the goals and means of resistance which are being applied today may well have been inconceivable a decade or two ago. As conditions change, people modify their perceptions and responses accordingly, and transform themselves into direct agents in effecting further change. The *intifada* is a prime example of this process of adjustment.

4 In many instances, women were actually fighting alongside men. They were also active in the 1936–1939 Revolt, performing such tasks as providing food and medical care to Palestinian fighters.

5 In what became the Occupied West Bank and Gaza Strip, charitable societies included orphanages, health programs, literacy programs, and other services and aid to the needy.

6 In the West Bank before 1967, peasant women worked alongside men in the fields to contribute to the subsistence economy of the peasant household. In 1967, such work accounted for some 64 per cent of the Arab female labor force; see A. Samed, 'Palestinian Women Entering the Proletariat', 1976: 159–67. By 1980, only 30 per cent of Palestinians earned their living from local agriculture; most Palestinian men went to work as cheap wage-earners in Israel; and Palestinian women were increasingly forced into seasonal work in Israeli agriculture. See Graham-Brown, 1984: 223–255; Samed, 1976: 159–167; and Palestinian–Jordanian Joint Committee, *Development of the Labor Force in the Occupied Territories*, Amman, Jordan, Palestinian-Jordanian Joint Committee, Publication No. 1, 1985 (in Arabic). Since 1967, subcontractors in the Israeli clothing industry have become a major employer of Palestinian women; see R. G. Siniora, *Palestinian Labor in a Dependent Economy: Women Workers in the West Bank Clothing Industry*, 1989.

7 Over the years, *In'ash* has greatly expanded its activities to encompass a number of training programs in sewing, canning, and marketing meant to enable women to become more independent and productive and to contribute both to the support of their families and to the development of their communities.

8 As official organizations, the charitable societies are organized under the rubric of the Union of Charitable Societies and are required to be registered with the Israeli authorities. This is necessary in order for them to receive licences to operate and to acquire permits for various activities and projects. In contrast, the Women's Committees, organized since 1989 under the broad mantle of the Women's Higher Council (which now includes a fifth distinct committee, to reflect the split within the DFLP into two separate factions), are not officially registered and are, therefore, liable to face restrictions in their activities or to be banned altogether at any time.

9 Apart from the women's committees, some of the more active grassroots committees include the Medical Relief Committees; the Voluntary Work Committees; the Agricultural Relief Committees; and the special youth groups, or *shebab*.

10 Essentially, the *intifada* aimed – at least during the first two years or so – to put pressure on Israel, by highlighting in Israel and abroad the issue of the continued occupation of Palestinian lands. Actions and tactics were directed, more

specifically, at making these areas ungovernable and at delinking, to the extent possible, from an almost total dependence on Israel. Behind the scenes of stone throwing and demonstrations, intensive work was being undertaken on all fronts to build Palestinian alternatives to Israeli institutions and to build the infrastructure of a future Palestinian state.

11 It is not possible to completely divorce a discussion of the role of women in the *intifada* from a discussion of the *intifada*'s overall goals, and the impact of the *intifada* – both real and perceived – on Israel. However, the present study concentrates mainly on the impact of the *intifada* within Palestinian society, particularly upon women.

12 This effort was largely uncoordinated and sporadic, and lasted, off and on, for about three years, until Israel began to allow universities to reopen. Similar efforts were attempted for secondary school students.

13 As a result of their participation and activism, women, too, were subjected to severe Israeli reprisals. They were beaten, arrested, even killed. Women suffered as a result of tear gas being thrown into their homes; there are, for example, no accurate statistics on miscarriages resulting from the effects of tear gas, but totals may well number in the hundreds. In addition, several dozen deaths can be attributed directly to tear gas; see PHRIC, *Monthly Updates*, and Al-Haq, *A Nation Under Siege*, 1990: 509–511. In other actions, the Israeli authorities broke into women's centers and ransacked Palestinian homes. In August 1988, Israel moved to ban all popular committees, see Al-Haq, *Punishing a Nation* 1989: 322. Though it is not clear that women's committees were included in the ban, this decision marked a turning point in the *intifada* and in the efforts of women within it.

14 By then, the PLO (during the Palestine National Council meeting of November 1988) had already declared an independent Palestinian state, recognized Israel, and renounced 'terrorism'; and yet no Israeli response was forthcoming. The 'dialogue' with the US, begun shortly thereafter, produced no tangible results, and instead seemed to marginalize the PLO even further. Meanwhile, Israel and the US were working on a scheme to impose limited autonomy under continued Israeli occupation in these areas.

15 In the early 1980s the Israelis were supplying these movements with arms and support and turning a blind eye to their activities. At the same time, the Israeli authorities were coming down hard on similar efforts to mobilize by secular groups; see for example, Taraki, 1989: 171–182, and her reference to a report in *The Jerusalem Post*, September 8, 1988.

16 The 'nationalist leadership' refers mainly to the Unified National Leadership of the Uprising. This body represents the main factions of the PLO and is responsible for charting the course of the *intifada* in the Occupied Territories. It was formed during the early weeks of the uprising in 1988. For more on the failure of the UNLU to adequately address the threat posed to women by Islamic fundamentalism, see Hammami 1990: 24–29.

17 Faisal Husseini, head of the Arab Studies Center in East Jerusalem and a prominent figure in the peace talks, made some remarks during the December 1990 Conference on the need for national consensus and respect for the role of women in the struggle; for a summary of Husseini's comments, see, Bisan Center, 'The Intifada and Some Women's Social Issues', April 1991: 6–7.

18 One need only look back at the acrimonious exchanges that took place between delegations of Western women, especially from the US, and 'Third World' women – including Palestinians – during the 1985 Nairobi Conference, that marked the end of the UN Decade for Women. During these encounters,

Palestinian women were quite firm in rejecting 'feminist' agendas. Such agendas, in their view, reduced women's problems to issues like day care, abortion and equal pay, when women were facing a threat to their very national existence. For more on the Nairobi conference, see Union of Women's Work Committees, 'Nairobi Forum: Women and Politics', 1985; A. Jou'beh, 'Women and Politics: Reflections from Nairobi', 1987: 50–57; and Dajani 1993. Essentially, both women from the Third World and women of color in the US challenge white middle-class hegemony over the terms of discourse, and insist on broadening the definitions of 'patriarchy' to include sensitivity to and analysis of issues of race, class, and ethnicity (including nationality), as well as gender. This effort to broaden the terms of discourse also introduces the need to study the 'nation-state' as a unit of analysis that has an impact on the oppression of women – again an area that has been so far largely neglected in much of Western feminist writings.

19 We may recall that between 1948 and 1967 the West Bank was ruled by Jordan. This area was declared formally annexed by Jordan in 1952, and Jordanian law extended there. Palestinians in the Occupied Territories continue to have very close ties to Jordan, and many envision a future in which the Palestinian state and Jordan are united in some kind of (con)federation.

20 Not discussed in this context, but worth noting, are the restrictions under which these women's centers operate. As in the case of the Women's Committees cited earlier, these centers are closely affiliated with one or another of the political factions of the PLO. Although some women activists claim to steer an independent course and to concentrate on investigating issues pertaining to women, most are constrained in their funding and operation and by their own factional loyalties. It is entirely possible, therefore, that some issues are toned down or avoided altogether, in order not to alienate the larger factional constituents and/or sources of funding. In all, it is difficult at this distance to do justice to such a complex issue. Suffice it to say that real constraints do exist, and that these constraints influence the intellectual products that emerge from these centers.

21 A number of texts in the feminist literature address this and related issues. In the US, for example, this contradiction is analyzed in terms of how race and gender are articulated with class, where women of the dominant white majority tend to discuss feminist concerns in a manner that excludes the realities of minority – especially African-American – women. This dichotomization pits white women (along with white males) against minorities. Similar concerns have been raised in Great Britain, both by white women and women of color there. See A. Gilliam, 'Women's Equality and National Liberation', 1991: 215–237; V. Amos and P. Parmar, 'Challenging Imperial Feminism', 1984: 3–21; and F. Anthias and N. Yuval-Davis, 'Contextualizing Feminism', 1983: 62–76.

22 For a comparative view of women and Islam in Iran, see N. Tohidi, 'Gender and Islamic fundamentalism', 1991: 251–271.

23 Israel, like the surrounding Arab countries, administers personal affairs for its citizens on the basis of religious law, and not on the basis of secular or civil codes. Israeli Jews are subject to Rabbinical law, and Muslim and Christian communities within Israel are ruled by their respective religious laws. Israeli Jewish women, therefore, devote much of their attention as feminists to these traditional patriarchal and religious structures.

24 One element missing in the case of South Africa is the Islamic religion. However, those who study the situation of black women in South Africa maintain that Christianity imposes its own brand of oppression upon women there; see, for example, C. Walker, *Women and Resistance in South Africa*, 1991: xii. See also D. Gaitskol and E. Unterhalter, 'Mothers of the Nation', 1989: 58–79. Much of

the information for this section is taken from Walker, 1991.
25 By 1902, Afrikaners in two colonies, the Transvaal and the Orange Free State, had already lost in their fight for independence against the British. Those two colonies, along with two others, the Cape and Natal, formed the new Union, Walker, 1991: 9.
26 Walker, 1991: 5. The state's efforts to preserve a traditional 'reproductive' role for women within the tribal economies collapsed under the weight of countervailing pressures to expand the pool of cheap labor. By the 1940s, black African women themselves were migrating to work in white areas, and the tribal social structures had begun to collapse. As the state was forced to use direct measures of coercion to maintain these reserves and protect its source of cheap labor, black women experienced directly the impact of Apartheid rule. It was then that a distinct women's movement began to crystallize around both national and social concerns; see Walker, 1991: 25.
27 In this decisiveness they may have been 'helped' by the fact that the South African government issued laws like the Pass Laws which particularly affected women. No comparable Israeli measures seem to have been targeted specifically against women, thus perhaps making it more difficult for Palestinian women to find a single external issue against which they could mobilize.

BIBLIOGRAPHY

Abdo, N. (1987) *Family, Women and Social Change in the Middle East: The Palestinian Case*, Toronto: Canadian Scholar's Press.
—— (1990) 'On nationalism and feminism, Palestinian women and the Intifada: no going back?', Paper presented at Roundtable on Identity Politics and Women, WIDER, Helsinki, 8–10 October.
—— (1991) 'Women of the Intifada: gender, class and national movement', *Race and Class*, 32, 4: 19–35.
Abdul Jawwad, I. (1990) 'The evolution of the political role of the Palestinian women's movement in the uprising', in M. C. Hudson (ed.) *The Palestinians: New Directions*, Washington, DC: Center for Contemporary Arab Studies, 63–77.
Abu Amr, Z. (1989) 'Nationalist and Islamic forces during the Intifada', Paper presented at Amman, Jordan: the Shoman Foundation, September.
Abu Ayyash, A. (1981) 'Israeli planning policy in the Occupied Territories', *Journal of Palestine Studies*, 11, 1: 111–124.
Al-Haq (1989) *Punishing a Nation. Human Rights Violations During the Palestinian Uprising, December 1987–December 1988*, Ramallah: Al-Haq.
—— (1990) *A Nation Under Siege. Al-Haq Annual Report on Human Rights in the Occupied Palestinian Territories, 1989*, Ramallah: Al-Haq.
Al-Khalili, G. (1977) *The Palestinian Woman and the Revolution*, Beirut: Palestine Research Center (in Arabic).
Amos, V. and Parmar, P. (1984) 'Challenging imperial feminism', *Feminist Review*, 17: 3–21.
Anthias, F. and Yuval-Davis, N. (1983) 'Contextualizing feminism: gender, ethnic and class divisions', *Feminist Review*, 15: 62–76.
Antonius, S. (1981) 'Fighting on two fronts: Palestinian women', *Al-Fajr*, March 8–14: 8–9.
Aruri, N. (ed.) (1984) *Occupation: Israel Over Palestine*, London and New Jersey: Zed Books.

Bisan Center (1991) 'The Intifada and some women's social issues', *Abstract of Conference held in Al-Quds/Jerusalem, 14 December, 1990*, Ramallah: Bisan Center.

Dajani, S. (1990) *The Intifada*, Amman, Jordan: Center for Hebraic Studies, The University of Jordan.

—— (1993) 'Palestinian women under Israeli occupation: implications for development', in J. Tucker (ed.) *Palestinian Women: Old Boundaries, New Frontiers*, Bloomington and Indianapolis: Indiana University Press and Center for Contemporary Arab Studies, Georgetown University.

Davies, M. (ed.) (1987) *Third World, Second Sex*, London and New Jersey: Zed Books.

Gaitskol, D. and Unterhalter, E. (1989) 'Mothers of the nation: a comparative analysis of nation, race and motherhood in Afrikaner nationalism and the African National Congress', in N. Yuval-Davis and F. Anthias (eds) *Woman–Nation–State*, London: Macmillan, 1989, 58–79.

Giacaman, R. (no date) 'Palestinian women and development in the occupied West Bank', Birzeit University, unpublished paper.

Giacaman, R. and Johnson, P. (1989) 'Building barricades and breaking barriers', in Z. Lockman and J. Beinin (eds) *Intifada: The Palestinian Uprising Against Israeli Occupation*, Boston: South End Press, A MERIP Book, 155–171.

Gilliam, A. (1991) 'Women's equality and national liberation', in C. Mohanty, A. Russo and L. Torres (eds) *Third World Women and the Politics of Feminism*, Bloomington and Indianapolis: Indiana University Press, 215–237.

Gluck, S.B. (1992) 'We will not be another Algeria: Palestinian women, nationalism and feminism', *Association for Women in Psychology, 1992 National Feminist Psychology Conference* (audio cassette).

Graham-Brown, S. (1984) 'Impact on the social structure of Palestinian society', in N. Aruri (ed.) *Occupation: Israel Over Palestine*, London and New Jersey: Zed Books, 223–255.

Hammami, R. (1990) 'Women, the Hijab and the Intifada', *Middle East Report*, 164–165, 24–29.

Hijab, N. (1988) *Womanpower: The Arab Debate on Women at Work*, Cambridge and New York: Cambridge University Press.

Hiltermann, J. (1990) 'Trade unions and women's committees: sustaining movement, creating space', *Middle East Report*, 164–165, 32–37.

Holt, M. (1992) 'Women in Palestine: going forwards or backwards?', *Middle East International*, August 7, 20–21.

Hudson, M. (ed.) (1990) *The Palestinians: New Directions*, Washington, DC: Center for Contemporary Arab Studies.

Hurwitz, D. (ed.) (1992) *Walking the Red Line: Israelis in Search of Justice for Palestinians*, Philadelphia: New Society Publishers.

Jad, I. (1990) 'From salons to popular committees: Palestinian women, 1919–1989', in J. Nassar and R. Heacock (eds) *Intifada: Palestine at the Crossroads*, New York: Praeger Publishers, 125–143.

Jammal, L. (1985) *Contributions by Palestinian Women to the National Struggle for Liberation*, Washington, DC: Middle East Public Relations.

Joseph, S. (1986) 'Women and politics in the Middle East', *The Middle East Report*, No. 138: 3–8.

Jou'beh, A. (1987) 'Women and politics: reflections from Nairobi', in M. Davies (ed.) *Third World, Second Sex*, London and New Jersey: Zed Books, 50–57.

Khreisheh, A. (1992) 'Interview, Mother and Leader' in O. Najjar (ed.) *Portraits of Palestinian Women*, Salt Lake City: University of Utah Press, 148–156.

Kuttab, E. (no date) No Title, unpublished paper, Ramallah: Bisan Research Center.
—— (1992) Presentation at the Union of Palestinian Women's Associations 7th Annual Conference, Chicago, IL, 22–24 May 1992, videocassette.
Legrain, J. F. (1990) 'The Islamic movement and the Intifada', in J. Nassar and R. Heacock (eds) *Intifada: Palestine at the Crossroads*, New York: Praeger Publishers, 175–191.
Lockman, Z. and Beinin, J. (eds) (1989) *Intifada: The Palestinian Uprising Against Israeli Occupation*, Boston: South End Press, A MERIP Book.
Mogannam, M. (1936) *The Arab Woman and the Palestine Problem*, London: Herbert Joseph.
Mohanty, C. (1991) 'Cartographies of struggle: third world women and the politics of feminism', in C. Mohanty, A. Russo and L. Torres (eds) *Third World Women and the Politics of Feminism*, Bloomington and Indianapolis: Indiana University Press, 1–5.
Mohanty, C., Russo, A., and Torres, L. (eds) (1991) *Third World Women and the Politics of Feminism*, Bloomington and Indianapolis: Indiana University Press.
Najjar, O. (1992) *Portraits of Palestinian Women*, Salt Lake City: University of Utah Press.
Nassar, J. and Heacock, R. (eds) (1990) *Intifada: Palestine at the Crossroads*, New York: Praeger Publishers.
New Outlook, 'Women in action', June/July 1989.
Palestine Human Rights Information Center (PHRIC) (1992) *Recreating East Jerusalem*, Jerusalem: PHRIC.
—— *Monthly Updates*.
Palestinian–Jordanian Joint Committee (1985) *Development of the Labor Force in the Occupied Territories*, Amman, Jordan: Palestinian–Jordanian Joint Committee, Publication No. 1 (in Arabic).
Samed, A. (1976) 'Palestinian women entering the proletariat', *Journal of Palestine Studies*, 16, 1; 159–167.
Sharoni, S. (1992) 'Middle East politics through feminist lenses: towards theorizing international relations from women's struggles', Paper presented at the Annual Meeting of International Studies Association, Atlanta.
Siniora, R. (1989) *Palestinian Labor in a Dependent Economy. Women Workers in the West Bank Clothing Industry*, Cairo, Egypt: Cairo Papers in the Social Science, 12, 3.
Strum, P. (1992) *The Women Are Marching: The Second Sex and the Palestinian Revolution*, New York: Lawrence Hill Books.
Taraki, L. (1989) 'The Islamic resistance movement in the Palestinian uprising', Z. Lockman and J. Beinin (eds) *Intifada: The Palestinian Uprising Against Israeli Occupation*, Boston: South End Press, A MERIP Book, 171–182.
—— (1990) 'The development of political consciousness among Palestinians in the occupied territories, 1967–1987', in J. Nassar and R. Heacock (eds) *Intifada: Palestine at the Crossroads*, New York: Praeger Publishers, 53–73.
Tohidi, N. (1991) 'Gender and Islamic fundamentalism: feminist politics in Iran', in C. Mohanty, A. Russo and L. Torres (eds) *Third World Women and the Politics of Feminism*, Bloomington and Indianapolis: Indiana University Press, 251–271.
Tucker, J. (ed.) (1993) *Palestinian Women: Old Boundaries, New Frontiers*, Bloomington and Indianapolis: Indiana University Press.
Tunis Symposium (1984) 'Women's committees in the occupied Palestinian territories and women's societies in the West Bank and Gaza Strip', *Conditions of Women's Organizations in the Arab Nation*, Tunis, December, p.8 (in Arabic).

Union of Women's Work Committees (1985) 'Nairobi Forum. Women and Politics', *Newsletter*.
Walker, C. (1991) *Women and Resistance in South Africa*, New York: Monthly Review Press.
Warnock, K. (1990) *Land Before Honor: Palestinian Women in the Occupied Territories*, New York: Monthly Review Press.
Women's Work Committees (1983) *The Internal Structure of the Women's Work Committees in the Occupied Territories*, n.a. (in Arabic).
Yuval-Davis, N. and Anthias, F. (eds) (1989) *Woman–Nation–State*, London: Macmillan.

4

HEIGHTENED PALESTINIAN NATIONALISM

Military occupation, repression, difference and gender

Tamar Mayer

Palestinian national consciousness has been developing steadily since the early years of the twentieth century, when the Arabs of Palestine actively resisted British policies favoring Jewish settlement there. More recently and dramatically, Palestinian national identity has been shaped in important ways by two major events: by the expulsion of Palestinians after the 1948 war and by the Israeli occupation of the West Bank and the Gaza Strip since 1967. It is because of territorial incursions – both geopolitical and domestic – which accompanied the Occupation of 1967, I will argue, that Palestinian people have become more politicized and more nationally active.

Land confiscation, military invasions of everyday life, daily harassment, curfews and arrests, shrinking employment opportunities and collective punishments have all been an integral part of the Israeli military occupation. The occupier/occupied power relationship which was imposed by Israel in order to control the indigenous Palestinian population and thus to 'ease the task' of occupation on Israel's part, has achieved only part of its goals, if any at all. The Occupation has proved unable to control fully the Palestinian people, or to control the Palestinian national spirit, especially as it has continued with no end in sight. The more Israel has tried to control Palestinian national feelings through harassment, arrests, and curfews, the more strongly these national sentiments have been expressed. The Palestinian national feelings brewing there before the Israeli occupation have intensified as a direct result of the Occupation, especially since the onset of the *intifada*. This attempt to 'shake off' the oppressive experience endured for more than twenty-six years was clearly the result of an existing national consciousness – but it also gave that consciousness an important push which has led to even more intense and proud national feelings among Palestinians. The Occupation gave birth to the *intifada*, and it has shaped the Palestinian national consciousness of today.

Although the literature on nationalism does not theorize about gender differences in the formation or expression of nationalism, the Palestinian case shows us that such differences do exist. Since Palestinian men and women have felt a similar attachment to the land, have experienced together the expulsion of 1948, the Occupation of 1967 and the recent *intifada*, it may seem that their feelings of nationalism would be formed similarly as well. But because Palestinian men's daily lives take place mainly within the public sphere while Palestinian women are largely confined to the private sphere, men and women also experience the Occupation in different ways. In particular, I will argue that the Occupation's attack on the private sphere is responsible in important ways for sharpening Palestinian women's national feelings and for crossing the line between public and private spheres. The Occupation's attacks on the family, on motherhood and on women's sexuality has incited Palestinian women to express themselves more militantly in the national struggle.

In this chapter, I will examine the role of the Israeli military occupation in heightening Palestinian national identity. I will examine the importance of territorial incursions in inspiring Palestinian national resistance and consciousness, and I will also show that the Occupation has intensified Palestinian nationalism in gendered ways by provoking a politicized response to the invasion of the private sphere. Ultimately, I will show that Israel's attempt to crush Palestinian nationalism has contributed actively to creating the enemy that Israel most feared.

NATIONALISM: CAUSES AND MEANS OF FORMATION

Even though nationalism is an historically specific phenomenon (Anderson 1988), most nations have developed their national consciousness in similar ways. If a group of people believe in a common origin and in the uniqueness of their history (even if it is mythical), and if they believe in a common destiny, share national symbols like language, religion, and customs, and have aspirations for a common territory (Smith 1981; Connor 1972), then we can assume that this group of people constitutes a nation. And since a nation is self-defined (Smith 1981; Connor 1978), it is the members of the nation who determine its existence. 'Nationalism,' then, is the form that motivates national behavior and mobilizes members of a nation to act.

The literature on nationalism also suggests that in addition to the importance of a common language, religion, customs, and a sense of a common past, destiny, and territory, attack on the territorial homeland or antagonism resulting from contacts with other peoples and cultures (Deutsch 1953) can also lead to the development of nationalistic feelings. Such antagonism, Hechter suggests, can occur especially when the nation sees itself as economically less developed, as when multinational states undergo modernization and economic development (Hechter 1975). In such cases, he argues, the development of the state usually comes at the expense of the national group, which is often exploited. The less economically developed a nation is, therefore, especially *vis-à-vis* other ethnic groups within the state, the more nationalistic its members are likely to become.

An examination of the Palestinian case will show that Palestinian nationalism has developed not only because Palestinians have been attached to their territory and because they have shared religion, language and customs and other nationalistic symbols, but because their history of expulsion, of dispossession and, more recently, the Israeli Occupation have sharpened and heightened their nationalistic feelings.

Palestinian nationalism

Because Palestinians share language, religion, and customs;[1] because they believe in a mythical as well as a real connection to their territory; and because they believe that they are a unique and separate nation, then they indeed constitute a nation, despite the fact that some may question its validity.[2] Palestinian nationalism and national consciousness have developed in significant part because of these people's blood connection and love for their land:[3] 'I am a Palestinian because I and my forefathers were born on this soil.'[4] Palestinian national feelings have been further stimulated by antagonisms which were exacerbated by the creation of a Jewish national homeland in Palestine (Joffé 1983) and later by the establishment of the state of Israel.

The Arabs of Palestine shared language, religion, and customs with the rest of the Arab world, and their own national consciousness was not initially formulated as independent of the rest of the Arab peoples. But the past fifty years separated them from their Arab brothers and sisters: their historical fate of dispossession and expulsion was not experienced by the rest of the Arab peoples; and the Palestinians felt abandoned by the Arab States in 1947, when Arab governments and armies rejected Palestinian pleas for economic and military support (Morris 1987). Yet while their developing sense of difference may, as Morris argues, have added to their sense of despair (1987: 17), and thus provoked the first buds of an independent consciousness among Palestinians, because most of the Arab population of Palestine did not have a distinct idea about statehood, Palestinian national consciousness remained relatively underdeveloped before 1948.

It was the fear and the actual displacement caused by Zionist Jewish incursions which substantially sharpened Palestinian nationalism and inspired Palestinians to fight, first against the British and then against the Jews. But Palestinian nationalism would not have grown to its present magnitude if it were not for the displacement of more than half a million Palestinians in the 1948 war with Israel and when the Arab states were defeated a second time, in the 1967 war. This last defeat, especially, marked a key change in Palestinian nationalism. For in 1967 it became much clearer to Palestinians that they could not rely on the Arab countries to deliver support for their national aspirations, and that the only way to achieve their national goals was to fight for themselves. Thus being a Palestinian has come to mean being dispossessed and, at the present, attempting to shake off the occupation which maintains that dispossession.

As Palestinian Arab land changed hands in the 1920s and 1930s and a sense of loss was building among Palestinian Arabs, the sense of connection to the land, to the community, and to the extended family (*hamoula*) which had been so characteristic of national consciousness in Palestine became increasingly more important. When in 1948 so many Palestinian Arabs fled and were expelled,[5] people left their homes believing that their displacement would be temporary: they took the keys to their houses (many of which were demolished by Israel); they took little money (Brand 1988); and while many took documents proving their ownership of the land (Grossman 1988), they also left clothes and other essentials behind. 'Because we left in a hurry and because I knew that we would be coming home soon, I left my house almost as was. I only took a little bit of clothes and my jewelry', said one of the women in my sample. Once their hopes of returning home began disintegrating, Palestinians' nationalism became anchored even more deeply in a sense of loss, not only of their lands and homes and of proximity to their extended family, but also of the traditional agrarian way of life that had helped shape Palestinian communal identity.

While the yearning to return home, to the houses and the land which they had left behind, certainly marked the Palestinians as different from the rest of the Arab world, it did not mobilize the masses into action. The yearnings, the sense of loss and the idea of the 'return' to their lands created a distinct Palestinian folklore and culture within which the poetry of Mahmoud Darwish, Ali Hashem Rasheed and others directly addressed the connection and tie to Palestine. Essential as folklore and culture are to nationalism, they cannot alone formulate a focused political agenda. The creation of the Palestinian National Liberation Movement (*Fatah*) in the late 1950s[6] made it possible for Palestinians to organize themselves to actively reverse their fate. Further, in a 1964 meeting in Jordanian Jerusalem of the Palestinian National Council, which served as a government-in-exile, the Palestinian Liberation Organization (PLO) was formed (Cobban 1984; Brand 1988). Yet while the existence of a government-in-exile type of organization attests to a high degree of national consciousness and national organization, this involvement was mostly limited to a small group of educated, middle-class people living in the Diaspora, outside Palestine itself. The majority of the Palestinians simply kept alive dreams about their homes and land, passing them on to the generation born after 1948 and nourishing the collective memory of loss.

It was not until 1967 that Palestinians as a larger whole started to clarify the distinction between themselves and the rest of the Arab world, as the uniqueness of their fate was highlighted. While Arabs all around them were living in separate sovereign states, Palestinians had no access to their homeland and most of that homeland had become occupied by Israel in 1967. At this point the Israeli Occupation – which marked Palestinians as separate from their independent fellow Arabs – came decisively to mark their separate nationalism.

Occupation and nationalism

The Israeli Occupation of the West Bank and the Gaza Strip which started in 1967 and is now in its twenty-seventh year has shaped Palestinian consciousness in unique and essential ways. Although of course Palestinian national folklore, language, religion and customs existed for generations before 1948, the Occupation and the *intifada* have cemented in blood Palestinians' common connection to territory and have enhanced and perpetuated Palestinians' sense of common history and their sense of difference from the rest of the Arab world. Ultimately, those aspects of Palestinian national identity which are rooted in fears of displacement have been strengthened by that very displacement and by the continuous occupation of Palestinian land by Israel.

The 1967 war and the additional loss of Palestinian land in the West Bank and Gaza to Israel sharpened Palestinians' sense of loss, expulsion, and dispossession. Even though the loss of land may not at that point have been enough to motivate Palestinians to act on their national feelings and attempt to free their lands from Israel, as had been the case after 1948, they were organized to do so by the PLO acting as the Palestinian government-in-exile. The PLO shifted its focus after 1967 from organizing the efforts of Palestinian exiles to return to their homes to mobilizing resistance among the Palestinian population under occupation (Cobban 1984). Still, because many West Bank Palestinians had initially hoped that their future would be determined by political negotiations between Israel and Jordan, relatively few of them participated at the outset in the full range of PLO initiatives.

It was actually the practices of the Occupation itself – the detention and deportation of activists, educators and leaders together with the lack of clarity about the duration of the Occupation – that eventually mobilized Palestinians to respond to PLO organizational initiatives. Since many of those deported by Israel in the late 1960s and early 1970s joined PLO activities in Jordan and Lebanon, the connection between the Palestinian Diaspora (especially its leadership) and the Palestinians of the Occupied Territories has been cemented, further facilitating the organization and mobilization of Palestinians. The struggle to weaken Israel – which was carried out by the PLO and other Palestinian groups[7] and whose success seemed earlier a virtual impossibility because of Israel's post-1967 military might – has become a national goal among Palestinians. Finally, Israel's handling of the military occupation has also yielded a stronger national consciousness among Palestinians: Israel initially had no clear plan for what to do with the West Bank; later, when Likud took power in 1977, it imposed on the indigenous population its vision of the West Bank as part of Biblical Israel and thus inseparable from Israel; and Israel has attempted to crush every Palestinian attempt to organize nationally in the Occupied Territories.

Nurtured on their grandparents' and parents' memories of the 1930s and 1940s, the younger generation of Palestinians under thirty has been born into the realities of the Occupation and is deeply conscious of the fact that they are the third generation of Palestinian Arabs to endure either occupation or dispossession, and sometimes both: It is 'because of the disaster of all Palestinian people, the expulsion of my parents and grandparents from their soil and the bombing of refugee camps that I feel so much stronger as a Palestinian.' Moreover, the relatively higher rates of literacy among Palestinians (higher on the average than the rest of the Arab world) has enabled more Palestinians to write about their experiences of the oppression of the Occupation and many more to read about these experiences (even while they personally may not have undergone them). During the late 1960s and 1970s the number of newspapers published by Palestinians in the Occupied Territories and their circulation increased. Newspaper stories about the Occupation and editorials condemning the Occupation began reaching even the most remote villages, integrating them into the Palestinian national struggle. Together with newly available radio and telephones, these instruments of mass communication have made the West Bank ever smaller, and this has promoted increasing unity of Palestinian national consciousness.

Relative economic deprivation

While the Israeli government has attempted to crush virtually every Palestinian effort to organize nationally, Israel has also taken advantage of the vast unskilled and semi-skilled labor pool which suddenly became available to her in 1967. Again, Hechter suggests that such an unequal economic relationship will result in heightened nationalism on the part of the economically more deprived group. Since this inequality dates back to the period of 1948 to 1967, relative economic deprivation has been shaping Palestinian nationalism since long before the Israeli Occupation of the West Bank and the Gaza Strip. But the Occupation has substantially intensified the nationalizing effects of economic inequities.

While Israel developed economically after 1948 and prepared to absorb thousands of new Jewish immigrants, the close to one million Palestinians who remained within the borders of the newly constituted Israeli state (Rubinstein 1990) were largely unskilled and uneducated, and thus were relegated to many of the menial jobs in the newly established Israeli economy. Moreover, while the Jewish towns and villages which sprang up throughout Israel – often the product of the labor of unskilled Israeli Arabs – were immediately connected to water, electricity and other infrastructures, Israeli Arab villages and towns did not experience similar infrastructural expansions, and were intentionally kept undeveloped. But because these Palestinians were under Israeli military governance, and because most of their leaders fled at the beginning of the war, they were not organized to resist Israeli rule. The Palestinians who fled to Jordan also served their host country economically at the expense of indigenous economic development in the West Bank (Benvenisti 1984). The West Bank was the fruit and vegetable basket of Jordan (Shadid 1988) producing 65 per cent of its vegetables, 60 per cent of its fruits, and 30 per cent of its cereals (Brand 1988: 157),[8] even though very little Jordanian money was invested there in industry or even in roads and other infrastructures. Although Hechter has theorized about the relationship of internal colonialism to the strengthening of national consciousness (Hechter 1975), the Palestinians in the West Bank remained quite dependent on Jordan and did not, at least initially, develop much autonomous nationalism.

But relative economic deprivation emerged as a major factor in the intensification of nationalistic feelings among Palestinians when it became intensely exacerbated by the Israeli military occupation of the West Bank and the Gaza Strip. Since no industrial infrastructure had been prepared between 1948 and 1967 in these areas and since over the years there had been a decline in local agricultural activity, major employment opportunities for Palestinians, the key to their economic survival, fell into the hands of their occupier. As a result of the war and Israel's physical expansion, the Israeli economy continued to develop rapidly and to exploit the cheap labor force which was now even more available in the Occupied Territories. Palestinians helped to build the new Jewish communities which were established in the West Bank (sometimes on confiscated Palestinian land), and they were integrated into the Israeli economy, within Israel, as unskilled and semi-skilled workers; at the same time, Israeli/Jewish companies became dependent on markets in the West Bank and the Gaza Strip for a large share of their sales. Israel sought both to take advantage of the Palestinian labor pool for its own development and to make sure that Palestinians were employed so that any social unrest could be prevented. At the same time, many Israeli leaders hoped that employing Palestinian men and women in the Israeli labor market would weaken the intensity of these workers' national feelings; the longer they were employed by Israel the more compromised their national identity would become (Mayer 1990). Not only have these Israeli hopes not materialized, but in fact the Israeli labor market has served as an arena within which Palestinians' awareness of inequities in their economic relationship with Israel and, in turn, of their own national feelings, have been heightened.

Palestinian participation in the Israeli labor market has increased steadily over the past twenty-seven years: While in the late 1960s the Israeli economy employed only a few thousand registered Palestinians, by 1984 close to 90,000 registered Palestinian workers were crossing the Green Line daily, and the numbers of registered and unregistered workers reached 110,000–120,000 by the late 1980s[9] (Shelley 1989). Although they have functioned as 'hewers of wood and drawers of water', Palestinians have participated in almost 80 per cent of the employment categories within Israel (Semyonov and Lewin-Epstein 1987) and constitute about 30 per cent to 40 per cent of the Israeli workforce (Shelley 1989: 33). But even though the Israeli economy has become extremely dependent on these Palestinian workers, Palestinians traveling to and from work in Israel continue to face harassment and these difficulties have actively provoked nationalist feelings among them.

Because Israeli laws ban Palestinian workers from staying overnight in Israel, the daily commute is a necessity. Therefore, in addition to the nine-hour days that many of these people work they often must drive several hours at each end of the work day. Their work days are further lengthened by investigations and searches which they face at checkpoints when crossing to and from Israel. These daily checks and searches have aroused consciousness among Palestinian workers of the differences between them and the Jewish residents of the West Bank – who travel to work in Israel and of course are never stopped by the Israeli authorities. Palestinian workers' sense of difference has, in turn, taken the form of a positive sense of nationalism: 'because the forces of the Occupation ask me every single day for my ID card I am reminded that I am a Palestinian'; or, in the words of another, 'because I am humiliated every single day by soldiers who investigate, search and question me at the roadblocks I know that I am different, that I am "the other" and that I am a proud Palestinian.' In addition to the workers who commute daily, there are those Palestinians for whom Israeli employers find illegal overnight accommodations. Although they avoid the lengthy trip home and the daily harassment at check points these Palestinian workers face other forms of humiliation and harassment which have also served to enhance, rather than to crush, their sense of national pride. They are often cramped several to a tiny room or to a garage, locked in for the night and unable either to travel freely around their neighborhood or to flee in an emergency.[10] In response, one such worker emphasizes: 'because I work in Israel I feel stronger about my Palestinian identity. Since I have come to work here I have confronted many problems that remind me daily that I am an oppressed and unwanted Palestinian. This encourages me to love my people and belong to the Palestinian people even more than ever before.'

Finally, most Palestinian workers have become increasingly aware of the fact that they have facilitated the development of their occupier's land and economy while receiving only minimal compensation. As they compare their economic status to the status of Israeli Jews, and consider how their own exploitation has contributed to that inequity, many Palestinians' sense of unfairness is expressed in terms of territory and nationhood: 'I work in a factory, I see that my land gives beautiful fruits which are exported to other countries and I get nothing from my own land'; 'I work for my enemy and improve his well-being, and what do I get in return? high taxes and humiliation'; or 'this is my soil, and the land of my forefathers and instead of working it for myself, I work for a state that is owned by others. It is the state of the enemy.'

Oppression, harassment, and the *intifada*

Palestinian nationalism has also been sharpened by the oppression, humiliation, and daily harassment which have characterized the prolonged Israeli occupation. Being occupied has meant being denied rights, being administratively detained,[11] and being punished collectively; living under curfews, with fear and violence; and constant subjection to Israeli military rule. The Israeli army has acted on its belief that attempts to organize and celebrate Palestinian national feelings are hostile and subversive to Israel, and thus need to be suppressed whenever possible, before they spread throughout the Occupied Territories. But, again, instead of crushing the Palestinian spirit, the occupier's repressive tactics have in large measure intensified national feelings among Palestinians: 'As the iron fist of occupation gets tougher my Palestinian identity gets a boost'; 'because the official and unofficial policy of Israel has been towards erasing our identity and erasing the Palestinian case, they have harassed, oppressed and humiliated us. Their brutality only makes me feel stronger as a Palestinian'; 'My Palestinian identity has grown stronger over the years, mainly because the Occupation has deepened our disasters and made me a lot more extreme in my national feelings than I was ever before.' Living with the invasive and long-lasting conditions of the Occupation and, at the same time, listening to the dreams, memories and stories of the elderly about the homeland has politicized many previously unpoliticized Palestinians and made them wish to bring an end to the Occupation. The most actively focused and broadly carried out attempts to achieve such change so far have been embodied in the Palestinian *intifada*.

Palestinians' attempts to shake off the Israeli Occupation and to shape their own fate have been at the center of the Palestinian *intifada*. It is important to note that without the existence of a well-organized grassroots national movement – quite distinct from the PLO – Palestinians would not have been able to rise so efficiently against the Occupation. Even though the uprising was initially sparked by a relatively minor incident,[12] within days it became an almost unanimous challenge to the Israeli presence (Cobban 1991) and extended under the Unified National Leadership (UNL) to rural and urban areas all across the Occupied Territories (Darweish 1989).

HEIGHTENED PALESTINIAN NATIONALISM

Palestinian resistance to the Israeli Occupation has involved both violent and non-violent acts. Aimed at lessening Palestinians' dependence on Israel and Israeli goods, the *intifada* has entailed boycotting Israeli goods, as well as refusing to work in Israel and to pay taxes to the Israeli government, as in the case of the West Bank village of Beit Sahour. Such non-violent acts have required the creation of a new economy and a new infrastructure within which home gardening, animal production, food distribution, garbage collection and health care provisions are intended to provide the necessary underpinnings of a successful rebellion (see Dajani, Chapter 3, in this volume for an extended discussion of women's roles in the creation of this new economy). Together with the throwing of stones and molotov cocktails and the burning of tires, these strategies have produced a new folklore, a new history with new heroes.

Because of its success in creating an alternative economy and reducing, at least temporarily, Palestinian dependence on Israel, the *intifada* has helped dramatically to boost Palestinian national unity, pride, and activism: 'the *intifada* has made me feel stronger'; '[it] has caused the fire of the revolution to grow and develop, and it united most Palestinians in order to achieve our independence.' Moreover, much of the daily 'disturbances' and the maintenance of the alternative economy have been carried out by youth whose more than 900 educational institutions (K-12 as well as universities) were closed down by military order because they ostensibly posed 'a threat to the security interests of the state of Israel'[13] (Usher 1991: 8). Palestinian adults have become proud of their youth's national commitment and have been strengthened by their rejection of passivity: 'our strength comes from the children: now I see the child who leaves his doll for a stone to be thrown at the armed soldiers. This child is sure that he and his stone are stronger than the soldiers. Fear is not important. And because the child has no fear, we can be proud of who we are.' At the same time Palestinian children's active participation in the national struggle reinforces in the next generation their own determination to shake off the Occupation. Thus the *intifada* has been both the expression and the agent of renewed Palestinian identity, strength and hope, boosted by the heavy focus on youth which has 'displayed a tone of assertion and affirmation as an articulation and expression of authenticity' (Ashrawi 1990: 81).

Once again, the deportations, arrests, and detentions intended by Israel to reduce Palestinians' national unity and pride have largely achieved the opposite effect.

In an attempt to crush Palestinian leadership in the Occupied Territories, and to control the indigenous population and the development of national feeling, the Israeli army has been deporting Palestinians, especially: journalists, newspaper editors, teachers, lawyers, student activists, and mayors, since the beginning of the Occupation. These deportations have become increasingly common since the beginning of the *intifada* and reached a dramatic climax with the deportation of more than 400 Palestinians in December of 1992. While the deportations were intended to create a vacuum in Palestinian leadership they have mainly helped to mobilize Palestinian leaders, primarily because many deportees immediately became involved in Palestinian organizations in the Diaspora. In addition, those who were left behind have continued the work of the deportees, and not allowed the vacuum desired by the Israeli authorities to develop.

In addition to deporting more than 1,500 Palestinians since the Occupation began, Israel has also arrested tens of thousands of Palestinians accused of 'disturbing order'; and the number of arrests has increased greatly since the beginning of the *intifada*. Among those arrested and detained have been youths throwing stones at the soldiers or people raising the Palestinian flag or handing out leaflets which spelled out regulations set by the leaders of the *intifada* (UNL), as well as people who were involved in acts of civil disobedience. Serving time in Israeli prisons and detention centers has, however, become a rite of passage for many Palestinian youths. At the notorious new detention center at Ketsiot, constructed to house the rising numbers of detainees from the territories, young Palestinians live in close quarters with thousands of others, many of whom are veterans in the Palestinian struggle. Young detainees serve time with educators, journalists and student activists who have lived, written and practiced resistance against the Israeli Occupation, and who are ready to discuss with them political philosophies and strategies. The commonality of their experience of Israeli repression has strengthened their national will and radicalized most of them: 'although they tried to break my spirit when I was in Ketsiot they could not, because we were all together, strong in our belief and love for our land and our people'; or 'my experiences in the Israeli prison crystallized my feelings as a Palestinian.' In a very real sense, then, the Israeli prison system has become one of the most effective schools for Palestinian nationalism.

In summary, Palestinian national consciousness has been shaped and sharpened by Palestinians' connection to the land of Palestine, by the experience of dispossession and expulsion and, most recently, by the last quarter of a century of Israeli occupation. While Palestinian nationalism had long been defined in opposition to the British and the Jews, it was, on one hand, the aftermath of the 1967 war, particularly the repressive features of the Occupation and, on the other, the resistance carried out by the *intifada*, that focused and sharpened Palestinians' feelings. The harsher the Israelis have been in their treatment of Palestinians, the more the Israelis have detained, deported, harassed and humiliated Palestinians, the more strongly many Palestinians have come to feel about their national cause: 'my Palestinian identity has increased dramatically because of the continual Israeli occupation, the murders, the breaking of bones, and because of the detentions.'

GENDER AND PALESTINIAN NATIONALISM

Palestinian nationalism has developed because of Palestinian Arabs' love for their land and family ties to the soil; in opposition to the British, the Jews, the Arab States and Israel; and in response to the experience of expulsions, dispossession and occupation – but not all Palestinians have experienced these events in the same way. In the past, for example, although class as a category of analysis did not distinguish differences in Palestinians' love for and connection to the land, it did distinguish among them in terms of how they experienced expulsion, dispossession, and the beginning of a new life in the Diaspora. The poor and the landless clearly fared a great deal worse under these conditions than did those belonging to the middle and upper middle classes.[14] Like class, gender also marks and creates key cultural, social, and political differences. This is not to suggest that gender differentiates in degree among Palestinians' attachment to the land or their love and commitment to the nation – but rather that because women's and men's daily experiences, including the experiences of expulsion, dispossession, and occupation, have been different, these forces may have sharpened men and women's nationalism in different ways from one another. Moreover, because historically men and women have played different roles in the national struggle, the perpetuation and enhancement of their nationalism seems likely to have been accomplished by different agents. In fact, such gendered differences do emerge on closer scrutiny.

Women and national consciousness: before 1967

While historically both Palestinian men and Palestinian women have participated in the national struggle, women's participation since the early twentieth century has been especially well documented (Hiltermann 1991; Peteet 1991; Warnock 1990; Kazi 1987; Hadad 1980). Women across the Third World have been involved, sometimes alongside their men and sometimes in separate women's groups, in resisting imperialism and foreign domination, on the one hand, and, on the other, feudal, traditional and exploitative local elites (Jayawardena 1986). Palestinian women's early struggles were, though, aimed almost exclusively against the British and the Jews.

Key differences between urban and rural women emerged quite early. While most Palestinian women did not participate in the armed struggle, at least not until the 1936–1939 revolts, urban women formed separate Arab women's charitable groups, like the Arab Ladies Association of Jerusalem (founded in 1919), Palestine Women's Union (1921), Arab Women's Society of Jaffa, and the Arab Women's Union of Jerusalem (1929)[15] which were also involved in the political struggle. Because these organizations were founded by urban, educated and upper class women who were family members of the leaders of the nationalist movement (Peteet 1991; Giacaman and Odeh 1988) and the labor movement (Hiltermann 1991), their agenda focused mostly on national rather than class issues, and was geared towards protesting British settlement policies and Zionist settlements (Hadad 1980). But because rural women were more severely affected by British colonial settlement policies and Jewish immigration than were urban middle-class women, since their access to land and thus to agriculture was threatened, their involvement in the national struggle was different. While urban middle-class women participated through over 200 charitable organizations (Brand 1988: 196), rural women participated through active demonstrations and bloody riots. Rural women's often militant, physical, confrontation of the British and the Jews eased urban women's involvement in the struggle while, at the same time, the charitable activities in which these urban women were involved focused on caring for orphaned victims of peasant rioters, the blind and the handicapped, and on educating mostly rural women (Sayigh 1989). Thus Palestinian women remained united despite differences in their immediate concerns and goals.

Because their national involvement started on grounds from which they were not challenging indigenous social structures – with literacy campaigns and aid to the needy, in the case of urban middle-class women and, in the case of peasant women, with demonstrations against the 'outsiders' – Palestinian women were able to take to the streets and move from the private to the public spheres in ways which were acceptable to their male counterparts. During the 1929 riots, for example, many peasant men were sentenced to death by hanging (Peteet 1991) and women took an important part in charging the atmosphere of their funerals with nationalistic mourning and meaning. While men became martyrs who died for the national cause, women became the mothers of martyrs, a position which would give them importance and visibility (also during the *intifada*) in the national struggle. Such activities took Palestinian women out of the isolation of home, family and community and out of their regional isolation, and made them part of the greater national struggle.

But the agenda for the Palestinian national struggle which was determined and acted upon in the public, political sphere never fully spelled out what would happen socially once the Arabs of Palestine successfully fought the British and the Jews. In the absence of a clear vision of statehood, for example, women's future remained located within the traditional realm even though through involvement in the national struggle they had become more visible in the public sphere. In fact, it seems that this was the most natural form of public activity for many Palestinian women, because they continued their involvement in such organizations through 1967 in the West Bank (under Jordanian rule) and the Gaza Strip (under Egyptian rule) (Hiltermann 1991).[16] While women's involvement in charitable organizations and demonstrations enhanced Palestinian national consciousness by facilitating ways to mobilize both women and men, to reach out to the more remote rural communities, and to raise women's literacy levels and their consciousness about the fate of the land, many of these women were not radical and we can probably assume that they felt comfortable in carrying out their national struggle through these traditional women's organizations, which in large measure preserved distinctions between private and public spheres. At the same time, while much of women's activities prior to 1948 were spontaneous and did not seek to mass mobilize the Arab women of Palestine (Peteet 1991) these organizations did pave the way for transition, for the later women's committees and the popular committees of the *intifada* would be based on their structure and message.

After 1948, urban women in the West Bank continued their involvement in charitable organizations: they cared for children and the wounded, and organized to ease the absorption of the expelled, especially those in refugee camps. But after the West Bank was annexed to Jordan in 1950, Palestinian women there yielded to Jordanian pressures not to organize politically. They dropped the adjective 'Palestinian' from the name of the Palestinian Arab Women's Society, devoted themselves fully to charitable activities, and withdrew from direct involvement in the national cause (Brand 1988). Because of Jordanian pressures any national activity or any attempt to raise Palestinian consciousness in which both men and women were involved had to be underground; and women's involvement in these underground organizations remained minimal.

But by the time the Israeli Occupation started, the level of national consciousness among Palestinian women of the West Bank had begun to rise. The PLO was established in 1964, and in 1965 the different Palestinian women's groups were united into the General Union of Palestinian Women (GUPW) which later became one of the many bases of the PLO (Sayigh 1989). GUPW became the primary, legitimated agency to represent Palestinian women and organize activities in all parts of the Palestinian Diaspora, from rural areas and refugee camps to towns and cities. By 1967, when the war broke out, Palestinian women were already well organized in activities which were aimed at raising their standard of living and their literacy rates and in military sessions which aimed at increasing Palestinian women's role in the national struggle (Brand 1988: 198). Their national involvement and sentiments would become ever more heightened as a result of the Occupation.

1967 to the present – the Occupation and national consciousness

Although both women and men have experienced the Occupation in many similar ways, their different roles in their own society and their location in either the public or the private sphere have also meant for men and women different kinds of exposure to the Occupation. Because more men than women travel to the Israeli labor market, for example, not as many women have experienced the same humiliation and harassment at the border or within the Green Line as men have – but at the same time they have confronted Israeli soldiers during the day, in their neighborhoods and homes, more often than Palestinian men generally have. In addition, since many Palestinian men have spent extended periods of time in detention camps and in the Israeli prison system, in a hotbed of Palestinian nationalism, many have come out of prison much more militant than they were when they went in. Even though women have also been detained and imprisoned for political activism, not as many women as men have 'graduated' from the Israeli prison system and its 'education' in Palestinian nationalism. And because the Israeli prison system is segregated by sex, even among those held for 'security' reasons, women have been unable to join men in the Palestinian national culture that develops there (al Hamdani 1987).

Perhaps most important, because the Occupation has attacked the private sphere, family and traditional life have been exposed and transformed, and important elements of motherhood, sexuality, and family relations have been shifted into the public arena, where power conflicts with the occupier have been carried out. In the Occupation's invasion of the private sphere, women have been affected more seriously than men, and therefore their national consciousness has been affected by agents which are different from those affecting men's national consciousness.

Family and motherhood

Because Palestinian society is patriarchal and sex roles have been so clearly defined in traditional ways, women were dependent on men for protection and maintenance and expected to be relegated almost entirely to the private sphere, to housework and to bringing up children (Darweish 1989). Although traditional societies undergoing modernization usually experience changes in the family, changes in the traditional Palestinian family have come particularly rapidly and intensely, especially in the context of the Occupation. The family and the home have become the turf on which much of the power struggle between the Israeli army and resisting Palestinians has taken place. The traditional Palestinian hierarchy has been altered and family affairs which used to be a private matter have become not just a community issue but often a major arena for the political and national discourse. As a result, the Palestinian family has become politicized and all its members, especially women and children, have entered the national struggle through it. As the Palestinian home has opened up, involuntarily, to the Israeli army and to its searches and violence, the home and the family have also become agents in the national struggle, increasing Palestinian women's nationalism.

Throughout the Occupation, and most severely during the *intifada*, the Palestinian home has been exposed and violated. In search of 'suspicious' or 'wanted' Palestinian young men, Israeli soldiers have entered homes, often in the middle of the night, awakened entire families, interrupted the privacy of men and women, searched every room, and humiliated and degraded parents in front of their children. The home has become yet another place where confrontations with soldiers take place. Such confrontations often elevate Palestinians' anger and hatred for the occupier to higher levels. This anger and hate are often, in turn, expressed in the streets in the days that follow a 'search' as women's involvement in the *intifada* and the involvement of children, which women do not attempt to stop, escalates.

In addition, houses of Palestinians involved in what the Israeli army believes to be subversive anti-Israel acts have been demolished, leaving thousands of Palestinian families homeless and without most of their household effects. Often in the bitter discussions with the Israeli soldiers on demolition duty Palestinian women have been central players even though their pleas have generally been to no avail. Also, because they spend more time in the home than men do, women tend to witness the demolitions and to be left with the task of restoring some kind of order to their household and family. Since for all Palestinian women house demolitions can never be justified, such acts have only deepened the hatred for the occupying army and strengthened these women's commitment to the Palestinian cause:[17] 'I was never political until they demolished my house and tried to ruin me and my family.'

At the same time as some Palestinian homes are being violated, either through searches or through complete demolitions, others serve as prisons. Whenever either men or women of any age are placed under house-arrest, and whenever there is a curfew the family home becomes the arena within which the sentence is carried out.

As the Israeli military has helped heighten Palestinian consciousness, it also helped erode the authority of the father, and by so doing helped transform the Palestinian traditional family. Because of the Occupation many Palestinian men have been absent from their families' lives, and that absence has threatened their authority within the Palestinian family. Palestinian women have in effect become heads of households in cases where the husbands have either been detained or expelled, or in cases where because of the shrinking agricultural opportunities men have been forced to seek work in the Israeli labor market. Employment across the Green Line means that whether men stay overnight in Israel or commute daily they cannot contribute to rearing the children and the responsibility of caring for the family, and maintaining its well-being rests almost solely on women's shoulders. In essence, males' absence from the household has pushed women into a more visible role in and out of the home which in many ways has helped raise their national consciousness.

As the private sphere has been invaded and has become in significant ways an extension of the public sphere, the Israeli military Occupation has also politicized the meaning of 'motherhood' for Palestinians. If before the Occupation 'motherhood' was confined to the home, the Occupation and particularly the *intifada* have given motherhood and mothering very public obligations. 'Motherhood' in the Palestinian context no longer means simply rearing children with stories about expulsion and dispossession of the past and biologically reproducing the next generation of Palestinians, or just building the social, psychological and economic well-being of children. Now 'motherhood' is also about politicizing children at an early age, supporting their national fight, defending them from Israeli soldiers and being visible in the streets. Palestinian women spend more time than men in the streets, confronting and sometimes negotiating with soldiers as a way of protecting their children. As they prevent Israeli soldiers from running after graffiti writers, Palestinian flag raisers, or even stone throwers, Palestinian women give the idea nationalized resonance: 'this is my duty as a mother to protect the children from the hands of the soldiers, from the enemy.'

Such political involvement with their children and in the streets has contributed to Palestinian women's national consciousness in ways which men have not experienced especially since Palestinian mothers have also become important symbols of nationalism, sacrifice and strength in their greatest moments of grief: during and after the funerals of children.[18] Many of these mothers who used to be apolitical have become politicized with the death of a child: 'if before my daughter's death I could only think about my work and about feeding her and her brothers, now I want to continue what she and the other *shaheeds* started. I want to hold unto this land ever harder . . .'; 'it is the blood of the *shaheeds* that paves my national path.' 'The mother of the *shaheed*' who since the early 1920s has lent support and strength to the mourning community of Palestine has continued to do so to this day. Her determination – 'we will continue to fight in the name of the *shaheeds* who died for Palestine' – only charges the atmosphere at the funerals[19] and at home with more national feelings and with renewed desire to continue the national fight.

Sexuality

With its harassment of women and security searches in homes and at the bridges,[20] the Occupation has also brought some issues of sexuality, which used to be confined to the private sphere, into the public sphere. Exposing women's sexuality to the public sphere has revealed the power structure not only between males and females but also between the occupier and the occupied. Women's sexuality has become an arena where these power struggles take place, but instead of intimidating these women and forcing them never again to participate in nationalist activities they have achieved the opposite: they made women more determined to fight their occupier and this, in turn, intensified their national feelings even more.

The presence of the Israeli army in Palestinian towns and villages has meant that women who prior to the Occupation were quite secluded now face men, mostly Israeli soldiers, in the public sphere. They now are forced to talk to men, to strangers, as they are asked for ID cards, or as they have to answer the questions of soldiers searching for 'wanted' Palestinians. They are often alone with the men who search their houses and who sometimes use this opportunity to molest, abuse and even attempt to rape them. Both young and old women have been forced out of their secluded life to encounter men in their streets and homes. This clearly threatens what it has meant to be a woman in the patriarchal Palestinian society. Moreover, as Palestinians cross the border from Jordan to the West Bank women as well as men are subject to army searches. Some such searches involve physical examinations in which Palestinians are forced to undress for the Israeli soldiers to make sure that they have not smuggled nationalistic literature or even ammunition. Finally, in an effort to discourage Palestinians from participating in any national activity, including participation in outlawed national organizations like the PLO, tens of thousands of Palestinians (mostly men but also women) have been detained, interrogated and sentenced to periods in prison. Many techniques have been used to force Palestinians' admission of their membership in these organizations – but Palestinian women have been the particular object of sexual harassment (Thornhill 1993). As part of the interrogation, Palestinian women's sexuality is often brought to the center by accusing them of being prostitutes and of being pregnant at the time of their imprisonment, so that their families will disown them.

For many of these women and their families sexual 'honor' is very important not only as a personal but, equally important, as a communal matter. Many of these Palestinian women suddenly find themselves alone with their male interrogators in a room.[21] While in their traditional society they were not allowed to talk to even one man now they have to talk to several of them and to endure vulgar discourse and frequent sexual innuendo. Yet although in the past these would be enough to bring 'shame' on a woman and her family, because the number of Palestinian women who have been exposed to such interrogators is growing these interrogations have, together with imprisonment, become a 'rite of passage' for many Palestinian women activists. By using Palestinian women's sexuality, intimidating and threatening them with 'shame', interrogators have hoped to use fear to make these women admit their participation in outlawed organizations: 'although I knew I was not pregnant, all I could think of was my father believing such made up stories.' In addition to threats about revealing women's 'past', interrogators threaten women with rape, and they often use a sexual language to get these women to admit their 'crimes':

> They beat me with a stick . . . on my stomach . . . [T]hen one day they told me to take my clothes off. This was the first thing I found myself unable to do. I was shaking. A big dark ugly man came in with a stick, and I knew he was going to rape me with it.
> 'Please don't! I'm young!'
> 'Are you going to take your clothes off, or am I?'
> 'OK', I said, and started to undress.
> 'You're not afraid?'
> 'No', I said, because I could see now that he wasn't going to do anything. His orders were just to frighten me. All eight interrogators came in, to fondle me, laugh at me, touch my breasts. I was so ashamed.
> (Warnock 1990: 150–151)

But instead of admitting their 'crime' these women often feel strengthened by their experiences: 'they tried to break me and my spirit but they did not . . . their interrogations made me so much tougher, they taught me a lot about my self and about them.'

Agriculture and work

The Occupation of 1967 meant, in addition to further losses of Palestinian land also a loss, for many Palestinians, of a traditional agrarian way of life. Because the West Bank could no longer provide most of Jordan's fruits and vegetables after 1967 (even though Israel has allowed limited agricultural export from the West Bank to Jordan as part of its 'open bridge policy'), and its agricultural produce could not be sold in the Israeli market for fear of undercutting prices for Israeli produce (Shadid 1988), changes in Palestinian agricultural production were inevitable. Moreover, because the Occupation expedited the process of modernization, Palestinian agriculture, like agriculture in many developing countries, has witnessed a radical decline in employment.[22] Many of the displaced men were forced to seek employment opportunities in Israel where, as for the most part daily laborers, they could no longer rely on a fixed income. Moreover, because many women who used to work in agriculture were displaced as well, they too, were forced to work outside their homes and support the family's economy, even while the traditional nature of Palestinian society in the West Bank has made most of them reluctant to travel freely to Israel and thus limited their ability to earn additional income. In the face of these major changes, Palestinian women are still making determined efforts to preserve elements of their traditional way of life. The changing conditions of agriculture and work under the Occupation have provided Palestinian women with another area in which to engage self-consciously in national acts of resistance, even though these acts may not change the fundamental structural realities of the situation.

The effective loss of subsistence farming and the pressures applied by the Occupation – including harsh limits set on Palestinians' access to water resources, deforestation carried out for security reasons, and restrictions on agricultural activities – have given working the fields new symbolic importance for Palestinian women; for it keeps alive their connection and commitment to the land and sustains their steadfastness (*sumud*) on their land, thus preserving Palestinian identity and heritage.

Threats to that 'steadfastness' posed by the Occupation and by the obligations to the Israeli labor market have been met with renewed commitment among many Palestinian women to maintain their hold on their lands for explicitly nationalistic reasons: 'the poor men, they have to work in Israel because there is no work for them in our country . . . we women, then, work on our pure and holy soil for them too, and this makes us feel more strongly the pride and the need to hold on to our land.' Even as with every day that passes farming is becoming more difficult for them, agricultural activity – which used to be one of women's major activities in the private sphere – has become an arena within which Palestinian women oppose the Israeli authorities and the Occupation's incursions into traditional Palestinian life and national identity, rooted in their connection to the land: 'I work this soil because I am connected to her and I love her . . . I will do all to make her bloom even if this means struggling with the occupying enemy.' Even more, women have been singled out by the leadership of the *intifada* for a major role in attempting to use 'family farming' to minimize Palestinian dependency on Israeli goods (Abdul Jawwad 1990: 72). Thus women who have traditionally maintained ties to the land have continued to do so during the uprising, but now around the heightened goal of national and economic resistance.

The loss of subsistence agriculture has also meant that Palestinian women must seek cash income, largely doing sewing under subcontracting to the Israeli labor market,[23] or as wage earners in Israeli factories within the West Bank, which specialize in finishing Israeli-made goods, especially in the garment industry (Talhami 1990). Here again, Palestinian women have sought out strategies for turning the threats to indigenous cultural survival posed by the Occupation into occasions for preserving their national heritage, especially through the sewing and embroidery of traditional Palestinian dress (Rishmawi 1988). Most Palestinian women have studied the art of sewing and embroidery at home as girls. Moreover, the colors, materials, and designs of traditional Palestinian dresses indicate village and regional identity (Sayigh 1992), so that in making traditional Palestinian clothes in order to increase the family income, women's preservation of particular village and regional designs and themes has also helped to develop national consciousness. At the same time, Palestinian children – who wear Western clothing like that of Israeli children since the West Bank is one of the larger markets for Israeli goods, including clothing – are exposed to traditional dress and learn the meaning of Palestinian themes in these women's sewing and embroidery. Finally, Palestinian women, like millions of women across the developing world, also work under sweatshop conditions. Their work disrupted traditional life and brought women out of the private sphere. In the Palestinian case, however, many of these potentially threatening changes to economic and family structures have been facilitated and monitored by different women's committees which have helped working women with child-care, health-care, literacy classes, and other activities which are intended to broaden these women's perception of their rights and to help them overcome their daily problems (Rishmawi 1988). Though aimed at helping women make the transition in a modern capitalist economy, all these activities have been aimed as well at politicizing women to strive not only for personal liberation but for national liberation, and to be conscious of the explicitly national needs which the threats posed by the Israeli Occupation have intensified.

CONCLUSION

Palestinian nationalism, like nationalism in other parts of the world, has been defined to a large extent in opposition to the 'other'. Historical events in Palestine and later in Israel made it clear to the indigenous Arab population that they, unlike other Arabs, were stateless, and dispossessed of their lands even if they shared language, religion and culture with the rest of the Arab world. Because much of their culture, heritage and folklore is anchored so heavily in their experiences of expulsion and dispossession, Palestinians' national identity has been defined in significant part by what the British, the Zionist settlers and, later, the state of Israel have done to them. Although Palestinians have resisted these historical events their resistance has at least not, yet, yielded much change for them – but it has clearly helped to build among them a sense of national identity and national mission.

The Israeli military Occupation, probably more than any other historical event, has dramatically sharpened Palestinian national feelings. At the same time as new generations of Palestinians have grown up with the Occupation, national memories of the past have remained very strong and have been nurtured, even among Palestinians born after 1948 and after 1967. The Occupation has since 1967 served as an increasingly unbearable reminder of the fact that being a Palestinian has meant being submitted to a fate determined by others. In response to the limits on their freedom and to the harassment and intimidation which have accompanied the Occupation since its beginning, Palestinian men and women rose in resistance in December 1987 to take their fate into their own hands: by creating alternative education and economic systems and by regularly challenging the Israeli military. Already a sign of high national awareness, the active resistance of the *intifada* has strengthened national identity and a positive sense of community among Palestinians, in part because it has attracted so much world attention and support: 'suddenly the whole world is talking about us and our case – and that makes me feel strong as a Palestinian.' At the same time the Occupation and the *intifada* have inspired a new type of national folklore, a new history and new heroes who together have instilled a renewed sense of confidence among Palestinians in the possibility of acting effectively in resistance against Israeli territorial threats to the Palestinian national homeland.

But while the memories of the past and the experiences of the present have influenced the development of Palestinian nationalism, men's and women's different daily experiences have also exposed them in differing ways to the Occupation and created gender difference in their attainment of nationalism. In particular, the Occupation's invasion of the private sphere of family, home, sexuality and even work – in essence, the forced entry of the public sphere into the private – has meant for women a different type of territorial incursion, equally essential to the formation of national feelings. As the Palestinian population has been pressured to work in the Israeli labor market by the increasingly limited employment opportunities in the Occupied Territories, men have encountered daily harassment when crossing the borders and in Israel, while women have not had the same kind of first-hand experiences in Israel itself. It has been their encounters with Israeli soldiers in their own neighborhoods, villages, and homes that intensified Palestinian women's national feelings. The Israeli soldiers who have ruthlessly entered the Palestinian private sphere have disrupted key aspects of family, motherhood and women's sexuality, intensifying women's anger at Israel, on the one hand, and, on the other, reinforcing their commitment to the justness of their own cause.

Most important, because Israel has constructed and treated occupied Palestinians as the enemy it has intensified nationalistic feelings on the part of the Palestinians – and in doing so directly contributed to the creation of the enemy that Israel feared most: self-conscious; proud; committed to resistance. Particularly in the case of Palestinian women, who were largely confined to the private sphere in traditional society, the Israeli Occupation has actively helped to politicize and radicalize its own opposition.

NOTES

1 Even though the Palestinian community consists of Muslims and Christians, the majority of the population is Muslim.
2 On June 15, 1969, Golda Meir, then Israel's prime minister, dismissed the existence of the Palestinian nation in a public address when she said, 'it is not as though there was a Palestinian people in Palestine considering itself a Palestinian people and we came along, threw them out, and took their country away from them. They did not exist' (*New York Times*, June 16, 1969).
3 Such strong connection to the land is a well-known phenomenon in pre-modern, mostly agrarian societies, where livelihood and culture are derived from and

anchored in the land. The Palestinian community's attachment to the land, therefore, is not unique.

4 All the quotes in this chapter not otherwise cited are taken from a field survey which I have carried out in the West Bank in 1989–1990. More than 500 Palestinians were interviewed: close to 300 men were commuters to the Israeli labor market, 150 males who worked in the Occupied Territories and over 100 women. Quotations from their responses will be presented throughout this chapter.

5 Israeli official record and the view of most Palestinians on expulsion of the Arabs of Palestine at the establishment of the Jewish state are quite different. Although many Palestinians left voluntarily because they believed that their departure from their homes and land would be short and temporary, others were clearly forced to leave. For excellent discussions of the creation of the Palestinian refugee problem see B. Morris, 'The Causes and Character of the Arab Exodus from Palestine: The Israel Defense Force Intelligence Branch Analysis of June 1948', in *Middle Eastern Studies*, 1986, vol. 22, no. 5, pp. 5–19; and B. Morris *The Birth of the Palestinian Problem*, 1987.

6 The first attempts to organize 'Palestine for its original inhabitants' took place in Cairo in 1951 when Y. Arafat organized a handful of Palestinian students who agreed with his ideology. In 1957 the Cairo student group, although still very small, started to disperse throughout the Arab world, where members met other displaced Palestinians, some of whom had already developed a similar ideological position. This ideology of returning Palestine to the Palestinians would later become the ideology of the *Fateh* and the PLO (Cobban 1984).

7 Although the PLO has been the sole representative of the Palestinian people and thus the highest authority on Palestinian nationalism, some dissenting groups have seen the PLO as too accommodating to Israel and other, more radical, groups have been fighting as the only way to achieve national sovereignty. These Palestinian groups have not disagreed on the principal goal but rather on the best means for achieving it.

8 In 1966 the West Bank contributed 45 per cent of Jordan's GNP.

9 These numbers have decreased since the beginning of the *intifada*, especially after Israel sealed off the West Bank and the Gaza Strip in April 1993 in response to the killings of thirteen Israelis by Palestinian workers; only about 40,000 Palestinians are currently permitted to work in Israel.

10 There have been several cases where Palestinians were housed in a garage or a small apartment and angry Jewish neighbors set fire to the 'housing unit' trapping and dooming to death the Palestinians whose employer had locked them in there without a key.

11 Palestinians are placed under administrative detention as charges against them are being prepared. The detention can vary in length from a week or so to a period of three or six months, and can also be extended indefinitely. Often the goal of these detentions is to limit the mobility and thus the activity of people who Israel believes are involved in 'subversive,' 'anti Israel' political work. Many of the detainees, especially since the beginning of the *intifada*, have been college students.

12 The *intifada* started in the Gaza Strip after the killing of four Palestinians and the injury of seven in a crash between a truck transporting the Palestinians from work in Israel and an Israeli military tank-transporter. The funerals in Jabaliyya refugee camp became the occasion for another massive demonstration against the Occupation, where the Israeli army teargassed and shot at the Palestinians, whose deaths produced yet a higher level of rage against the Israelis (Peretz 1990: 39).

13 The schools were closed by Israeli authorities, many believe, not only for security reasons but also in order to create a situation whereby the level of Palestinian education which has traditionally been very high would be lowered, enabling Israel to exploit Palestinians as hewers of wood and drawers of water and hampering the growth of the next generation of Palestinians.
14 People of the middle and upper middle class who were able to flee with money settled in towns and were often able to take on the same urban professions, while the rural population who fled or were expelled and who ended up in refugee camps were isolated not only from homes but also from their way of life. Their economic survival was a great deal more difficult than that of the middle and upper middle class Palestinians.
15 These women's organizations promoted literacy for girls and women, helped the needy and orphaned children, and provided literacy and sewing classes for women.
16 After 1948 there were no records of activities directly organized by Arab women in Israel (Hadad 1980: 160).
17 For an excellent account of the politicization of Palestinian women in light of house demolitions and to youths' political activity, see the interview with Salima Kumsiyya of Beit Sahour, in O. Najjar (ed.) *Portraits of Palestinian Women*, 1992: 105–119.
18 While many more men died in their national struggle, women too have sacrificed their lives and have been elevated to the status of *shaheed*. The first woman *shaheed* of the *intifada* was Haniyyeh Ghazawneh, who was killed less than a month after the *intifada* started, on January 3, 1988, as she was trying to rescue a child from being beaten by an Israeli soldier.
19 Because the Israeli army tries to maintain these funerals as family events and to prevent their escalation into nationalistic demonstrations, the army has often made families bury their dead in the middle of the night. This is especially so at times of curfew. This peculiar timing, however, usually does not deter the community from participating in the mourning, and from creating mock funerals for the *shaheeds* (during the day, hours and sometimes days after the person was physically buried) creating precisely what the army fears – yet another nationally charged event.
20 Palestinians who want to enter the West Bank from Jordan do so through the bridges over the Jordan river.
21 Even though Israeli security law requires the presence of a woman at all interrogations of women, when Palestinian women are interrogated this law is frequently not followed and Palestinian women often find themselves totally at the mercy of their male interrogators, who can do to them whatever they feel like.
22 The size of the labor force employed in West Bank agriculture has dropped in relative terms from 45 per cent of total employed laborers in 1969 to 19 per cent in 1984 (Awartani 1988: 143).
23 Women have constituted 45 per cent of the total agricultural labor force in the West Bank (Talhami 1990).

BIBLIOGRAPHY

Abdo, N. (1991) 'Women of the *intifada*: gender, class and national liberation', *Race and Class*, 32, 4: 19–34.
Abdul Jawwad, I. (1990) 'The evolution of the political role of the Palestinian women's movement in the uprising', in M. Hudson (ed.) *The Palestinians: New*

Directions, Washington DC: Center for Contemporary Arab Studies, Georgetown University, 63–76.

al-Hamdani, L. (1987) 'A Palestinian woman in prison', in Khamsin collective (eds) *Women in the Middle East*, London and New Jersey: Zed Books, 40–59.

Anderson, B. (1982) *Imagined Communities: Reflections on the Origin and Spread of Nationalism*, 2nd edn 1991, London and New York: Verso.

Anderson, J. (1988) 'Nationalist ideology and territory', in R. J. Johnston, D. B. Knight and E. Kofman (eds) *Nationalism, Self-Determination and Political Geography*, London: Croom Helm, 18–39.

Ashrawi, H. (1990) 'The politics of cultural revival', in M. Hudson (ed.) *The Palestinians: New Directions*, Washington, DC: Center for Contemporary Arab Studies, George Washington University, 77–86.

Awartani, H. (1988) 'Agricultural development and policies in the West Bank and Gaza', in G. Abed (ed.) *The Palestinian Economy: Studies in Development under Prolonged Occupation*, London and New York: Routledge, 139–164.

Benvenisti, M. (1984) *West Bank Data Base Project: A Survey of Israel's Policies*, Washington DC: American Enterprise Institute.

Brand, L. (1988) *Palestinians in the Arab World: Institution Building and the Search for State*, New York: Columbia University Press.

Cobban, H. (1984) *The Palestinian Liberation Organization: People, Power and Politics*, Cambridge and New York: Cambridge University Press.

—— (1991) 'The PLO and the *Intifada*', in R. Freedman (ed.) *The Intifada: Its Impact on Israel, the Arab World, and the Superpowers*, Miami: Florida International University Press, 70–106.

Connor, W. (1972) 'Nation building or nation destroying?', *World Politics*, 24, 319–355.

—— (1978) 'A nation is a nation, is a state, is an ethnic group, is a . . .', *Ethnic and Racial Studies*, 1, 4: 377–400.

—— (1984) 'Eco – or ethno – nationalism', *Ethnic and Racial Studies*, 7, 3: 342–359.

Darweish, M. (1989) 'The *intifada*: social change', *Race and Class*. 31, 2: 47–61.

Deutsch, K. (1953) *Nationalism and Social Communication: An Inquiry into the Foundations of Nationality*, Cambridge, MA: Technology Press of MIT.

Giacaman, R. and Odeh, M. (1988) 'Palestinian women's movement in the Israeli-occupied West Bank and Gaza Strip', in N. Toubia (ed.) *Women of the Arab World: The Coming Challenge*, London and New Jersey: Zed Books, 57–68.

Grossman, D. (1988) *The Yellow Wind*, New York: Farrar, Straus, Giroux.

Hadad, Y. (1980) 'Palestinian women: patterns of legitimation and domination', in K. Nakhle and E. Zureik (eds) *The Sociology of the Palestinians*, London: Croom Helm, 147–175.

Hechter, M. (1975) *Internal Colonialism: The Celtic Fringe in British National Development 1536–1966*, Berkeley and Los Angeles: University of California Press.

Hiltermann, J. (1991) *Behind the Intifada: Labor and Women's Movements in the Occupied Territories*, Princeton: Princeton University Press.

Jayawardena, K. (1986) *Feminism and Nationalism in the Third World*, London and New York: Zed Books.

Joffé, E. (1983) 'Arab nationalism and Palestine', *Journal of Peace Research* 20, 2: 157–170.

Kazi, H. (1987) 'Palestinian women and the national liberation movement: a social perspective', in Khamsin Collective (eds) *Women of the Middle East*, London and New Jersey: Zed Books, 26–39

Mayer, T. (1990) 'Nationalism and work: the case of Palestinians laborers in Israel', Paper presented at the Annual Meetings of the Association of American Geographers, Toronto, Canada, April 19–22.

Morris, B. (1987) *The Birth of the Palestinian Refugee Problem, 1947–1949*, Cambridge and New York: Cambridge University Press.
Najjar, O. (1992) *Portraits of Palestinian Women*, Salt Lake City: University of Utah Press.
Peretz, D. (1990) *Intifada: The Palestinian Uprising*, Boulder: Westview Press.
Peteet, J. (1991) *Gender in Crisis: Women and the Palestinian Resistance Movement*, New York: Columbia University Press.
Rishmawi, M. (1988) 'The legal status of Palestinian women in the occupied territories', in N. Toubia (ed.) *Women of the Middle East: The Coming Challenge*, London and New Jersey: Zed Book, 79–92.
Rubinstein, D. (1990) *The Fig Tree Embraces*, Jerusalem: Keter Publishing House, (in Hebrew).
Sayigh, R. (1989) 'Palestinian women: triple burden, single struggle', in Khamsin Collective (ed.) *Palestine: Profile of an Occupation*, London and New Jersey: Zed Books, 153–177.
—— (1992) 'Palestinian women: a case of neglect', in O.Najjar with K. Warnock (eds) *Portraits of Palestinian Women*, Salt Lake City: University of Utah Press, 1–26.
Semyonov, M., and Lewin-Epstein, N. (1987) *Hewers of Wood and Drawers of Water: Noncitizen Arabs in the Israeli Labor Market*, Ithaca., NY: ILR Press, Cornell University.
Shadid, M. (1988) 'Israeli policy towards economic development in the West Bank and Gaza', in G. Abed (ed.) *The Palestinian Economy: Studies in Development under Prolonged Occupation*, London and New York: Routledge, 121–138.
Shelley, T. (1989) 'Palestinian migrant workers in Israel: from repression to rebellion', in Khamsin Collective (eds) *Palestine: Profile of an Occupation*, London and New Jersey: Zed Books, 32–56.
Smith, A. D. (1981) 'State and homelands: the social and geopolitical implications of national territory', *Millenium: Journal of International Studies*, 10, 3: 187–202.
Talhami, G. (1990) 'Women under occupation: the great transformation', in S. Sabbagh and G. Talhami (eds) *Images and Reality: Palestinian Women under Occupation and in the Diaspora*, Washington DC: IAWS Monograph Series, Women of Palestine, No. 1, Institute for Arab Women's Studies, Inc,15–28.
Thornhill, T. (1993) 'The interogation of women "security" detainees by the Israeli General Security Service', in H. Afshar (ed.) *Women in the Middle East: Perceptions, Realities and Struggles for Liberation*, New York: St. Martin's Press, 185–205.
Usher, G. (1991) 'Children of Palestine', *Race and Class*, 32, 4: 1–18.
Warnock, K. (1990) *Land Before Honour: Palestinian Women in the Occupied Territories*, New York: Monthly Review Press.

5

ISRAELI WOMEN AGAINST THE OCCUPATION
Political growth and the persistence of ideology[1]

Yvonne Deutsch
(translated by Andre Rosenthal)

> By peace we mean the absence of violence in any given society, both internal and external, direct and indirect. We further mean the non-violent results of equality of rights, by which every member of the society, through nonviolent means, participates equally in decisional power which regulates it, and the distribution of the resources which sustain it.
>
> (Brock-Utne 1985: 2)

The Palestinian *intifada*, which began in December 1987, the 21st year of the Israeli occupation of the West Bank and the Gaza Strip, has promoted the political involvement of women in Israel against the Occupation. Close to a month after the beginning of the *intifada*, non-Zionist, radical, Jewish women in Israel began demonstrating in support of the Palestinian struggle against the Occupation, and eventually they succeeded in creating the political tools to involve together both radicals and women from the political center. Gradually more and more women, both Jewish and Palestinian, from Israel joined protest and solidarity activities, including meetings with Palestinian women from the Occupied Territories. Up until the crisis in the Gulf in August 1990 the women's peace movement was steadily developing in Israel – simultaneously mounting significant and widespread opposition to the Occupation and, at the same time, attempting to create an alternative political culture as the basis for radical change in the basic priorities of Israeli society.

The 1990 Gulf crisis and war constituted a turning point which brought a drastic drop in the scope of women's activities against the Occupation and in support of peace. Moreover, the new government of Israel, elected in June 1992, in which the liberal factions of the peace movement became part of the ruling Labor establishment, together with the political process shaped by the peace talks – which until the mutual recognition of Israel and PLO in September 1993 denied Palestinian rights to national autonomy – impeded the continued development of the women's peace movement. Still today, some five years after the beginning of the *intifada*, a core of Israeli Jewish women[2] continue, although on a reduced scale, their political activity in favor of peace. Today women from the Israeli political center fulfill the role previously played by radical women, attempting to create ties with Palestinian women and to conduct joint activities, while radical women now seek to redefine their role as a catalyst in creating changes in the political consciousness of women. In particular, feminists are trying to widen the scope of the feminist critique of war as a means for solving problems and of the relationship between, on the one hand, the priority of national security and the central role of the army and, on the other, the influence of militaristic culture on the status of women in Israeli society. The new developments of mutual recognition between Israel and the PLO may activate women, once again, to join political activities. The kind of activity now needed is one which will help pave the road to a mutually acceptable solution and reconciliation in the specific context of a feminist peace movement and political culture.

At the same time, the Israeli women's peace movement continues to act within the social norms of patriarchal society and to accept the notion that states depend legitimately upon military power for their existence and development. The basic concept of national security remains of special importance to the Jewish people because of historical precedents. Within the Israeli context, therefore, every feminist attempt to reexamine basic social norms and to position national security *vis-à-vis* personal security, especially of women, is likely to remain marginal even among women activists who work for peace.

ISRAELI WOMEN DURING THE *INTIFADA*: HISTORICAL DEVELOPMENT

The Palestinian *intifada* has clearly served as a catalyst for the political awakening of women in Israel, provoking them to act against the Occupation and for peace. Some measure of this awakening can be taken in the fact that within two years of the beginning of the *intifada*, in December of 1990, some 6,000 Israeli and Palestinian women, and women from Europe and the USA participated in a march for peace in Jerusalem and demanded that the Israeli government hold peace talks with the PLO, agree to an international peace conference, and permit the establishment of a Palestinian state alongside Israel.

This dramatic political event ended the 'Women go for peace' day which was organized by the women of the Israeli Women and Peace coalition, together with three Palestinian women's groups – the Federation of

Palestinian Women's Action Committees; the Union of Palestinian Working Women's Committees; and the Women's Committee for Social Work – along with women from the Italian Peace Association, *Casa delle Donne* and *Centro di Documentazione Donne*.[3] Throughout the day an international conference in favor of peace and a large demonstration by Women in Black were held in West Jerusalem. Organizers pinned their hopes on dialogue and joint activities between Israeli and Palestinian women as a basis for improving relations between the two peoples and for creating a strong women's movement in the Middle East.

This day was the climax of Israeli women's activity against the Occupation in exclusively women's groups which had begun about a month after the outbreak of the *intifada* and continued with ever-increasing intensity.

Women in Black

Dressed in black, in non-violent opposition against the Occupation, some ten Israeli Jewish women from the radical left began in Jerusalem in January 1988 to stand in solidarity with the Palestinian people. By making their political statement on the sidewalk of a main city square, surrounded by bystanders, these women exposed themselves to frequent outbursts of animosity and to an atmosphere of violence. Determined to continue, they transferred their protest site to Jerusalem's France Square, which was less exposed to interfering bystanders but was a route used by thousands of cars. Women in Black demonstrations spread to Tel-Aviv[4] and Haifa, and from there to kibbutzim, moshavim, and other villages around the country. By July 1990 participants counted some thirty women's vigils involving both Jews and Palestinians from Israel who demonstrated in black against the Occupation every Friday, at the same places and times, often to cries of 'whores!' 'traitors!', 'Arafat's whores!' and similar expressions.

As the popularity of Women in Black grew throughout Israel, these protest groups provided a model for women's protest groups in other parts of the world. Within a short space of time women in the USA, Italy, the Netherlands, Australia, Germany, and England had organized similar solidarity vigils. European women adopted this model for demonstrations against the Gulf War; and in the former Yugoslavia groups of Women in Black used it to organize against the civil war in their country and against the widespread phenomenon of women's rape. In India women have adopted this model to protest against the rape of women by religious extremists and against religious motives for violence toward women; and in Italy women adopted it to protest against the violence of the Mafia.

Shani: Women against Occupation

Founded in Jerusalem in January 1988, this group provided Israeli Jewish women with a framework for political education through house meetings with Palestinian women and with experts in the social and political spheres. *Shani* organized visits to women's groups in the Occupied Territories, which provided many Israeli Jewish women with their first opportunity to meet Palestinian women and to discuss with them controversial topics while searching for a joint political solution.

Women for Women Political Prisoners

Founded in May 1988, first in Tel-Aviv and then later in Jerusalem, this group provides practical and legal aid to Palestinian women political prisoners, establishes contact between prisoners and their families, and publishes information concerning conditions of imprisonment.

The Peace Quilt

Soon after the beginning of the *intifada*, thousands of Israeli women throughout the country began sewing a Peace Quilt composed of thousands of pieces of material decorated with drawings, writing, and embroidery expressing the desire for an end to the Occupation and for peace. In June 1988 hundreds of Jewish and Palestinian women participated in the unveiling of the quilt in front of the gates of the Knesset in Jerusalem.

THE CREATION OF A NATIONAL COALITION: WOMEN AND PEACE

To feminist activists, it was clear that in order to translate women's political awakening into a political force with real influence in Israeli society, a statewide movement of women needed to be created. In December 1988, at the end of the first year of the *intifada*, and as a result of the conference 'A call for peace: feminist perspectives on the occupation', the first nation-wide coalition of Israeli women in favor of peace, Women and Peace, was established.[5]

Limits on their power to influence political reality forced Israeli women to come to terms with their marginality in the decision-making process of the political establishment and in society as a whole. Their feelings of frustration and anger increased as they confronted the gap between the reality projected by decision makers and the values, beliefs, and life perspectives of women searching for peace. Since the road to influence within the existing political establishment was barred, these women chose the alternative of influence through independent, extra-parliamentary frameworks.[6] These political frameworks served in turn as catalysts and models for another group of women, Israel Women's Peace Net (*Reshet*). This group was founded in 1989 at the height of joint preparatory work on the event 'Women Go For Peace' between women of the coalition Women and Peace and Palestinian women. *Reshet*, led by women from the political center and Israeli academics, began holding house meetings and meetings between Palestinian women (middle to upper class; also mainly academic) and Israeli women who held positions of strength in the established women's groups and in Israeli centrist and leftist political parties (Chazan 1991). During the third year of the *intifada*, women of *Reshet* joined with Women and Peace and Palestinian women to formulate a peace treaty between Israelis and Palestinians based on feminist principles (Espanioly and Sachs 1991).

It is important to note that two factors contributed to success in the creation of an Israeli women's peace movement: the declaration of an independent Palestinian state in November 1988; and the meaningful development of a Palestinian women's movement during the first years of the *intifada*. From the Palestinian perspective the declaration of 1988 was regarded as a compromise, because it included acceptance of the partition of the land as part of a policy aimed at convincing the Israeli public of Palestinians' willingness to live in peace alongside Israel. At that time there prevailed a feeling of optimism among many Palestinians, due to what they saw as the success of the *intifada* in establishing a viable political organ capable of creating cultural and economic independence. This optimism also aided Israeli women's political activism on behalf of peace and strengthened relationships between women in Israel and the Palestinian women of the Occupied Territories. The peace march of December 1989, for example, was one of the outcomes of this development.

WOMEN'S MOVEMENT IN CRISIS: THE GULF WAR

The Gulf crisis in August 1990 found Israeli women activists in the midst of preparations for two conferences (one in Jerusalem and one in Geneva) which intended to open up dialogue with women of the PLO and to formulate with them the feminist peace treaty. But when the war broke out in January 1991, Jewish and Palestinian women found themselves on opposite sides. The majority of Jewish women in Israel felt threatened by the military might of Saddam Hussein and hence felt relief when the US attacked Iraq. Women and Peace was the only Israeli women's group which organized a demonstration against the war. The difference in Palestinian and Jewish women's political responses to the war stemmed mostly from their different identities and reference points: while most Israeli Jewish women have identified themselves with the West, in general, and with the US in particular, Palestinian women's point of reference has usually been the Third

World, and more specifically the Arab world. With the outbreak of the Gulf War women activists in Israel were quiet for a period of about three weeks. During the third and fourth weeks some women returned to their vigils in Jerusalem, Tel-Aviv, and Haifa; but of the thirty vigils which had been going on throughout the country before the war only a small number of much-diminished demonstrations survived.

The drop in activity against the Occupation was common among all Israeli peace groups; the less radical elements within these groups justified the change by pointing to their disappointment with the PLO's support for Iraq. Among many women there was also a deeper feeling of weariness with the ongoing vigil and political activity and the lack of results. In the early period of the *intifada*, activity against the Occupation and joint meetings with Palestinian women had provided a strong sense that a real alternative to the prevailing relationship between the two peoples was being created and that movement toward an acceptable political solution and peaceful coexistence in the region was taking place. But the war threw Israeli women back into frustration with the growing gap between their stance in favor of peace and the lived reality of ongoing war that was exacerbated by their distance from and inability to influence either the US and its allies or Iraq.

Palestinian women were also forced to cope with feelings of disappointment: with their inability to break through the political barriers of Jewish society; with the deepening of the existential threat to their lives under the Occupation; and with their inability to see any political breakthrough from which to gain hope. Political despair among Palestinians generally and in the Territories specifically has led steadily to an increase in violence, the strengthening of Islamic fundamentalist movements, and the erosion of the social and political position of women.

THE NEW ISRAELI GOVERNMENT

The 1992 Labor Party victory in Israel and the new coalition government brought to many liberal Israeli peace activists a temporary gleam of hope about the prospects for real political change. After September 13, 1993 and the signing of the interim peace agreement between Israel and the PLO – which was preceded by mutual recognition – there is indeed a real prospect for peace. It must be borne in mind, however, that the same government which signed this peace agreement deported over 400 Palestinians from the Occupied Territories in December 1992; sealed the borders between the Occupied Territories and Israel in April 1993, without first establishing an infrastructure capable of providing for the large number of Palestinians who had previously found work in Israel; and waged war in Lebanon in July 1993, forcing the mass movement of a civilian population, to serve political motives.

The Zionist parts of the peace movement were partners in these acts, and therefore only a few demonstrations opposed them. These political changes – the change in government; the peace talks (which have both raised hopes and disappointed, because it is not clear that they are leading towards a recognition of Palestinian national rights); and the worsening of living conditions in the Occupied Territories – have also created embarrassment and stimulated the search for appropriate action among women's groups.

THE NEW FEMINIST AGENDA

Taking into consideration these changes, the developments within the coalition Women and Peace are of particular importance. To a certain extent the coalition is today at a political crossroads and needs to redefine its role in the light of the fact that women from the political center are now undertaking political activity which was up until a short while ago delegated only to the radical left. Women of the Israel Women's Peace Net (*Reshet*) are working on advocating Palestinian human rights and looking for ways to renew political relationships with Palestinian women. The active women of the coalition, who are on the whole the organizers of Women in Black throughout the country, dedicate their time to the following three main areas.

First, as stated above, during the war the voices of dissent were quiet and a true awakening had not yet taken place. Nonetheless it was clear that the level of violence against women had increased significantly in Israel and that the number of murders of women by their abusive spouses had multiplied. The feminist activists of the coalition, who dedicated much of their time to demonstrating against the Occupation, found themselves devoting most of their energy to activities of traditional feminism such as creating public accountability for violence against women.

Second, traditional leftist women continued to search for forums for joint demonstrations with Palestinian women and even joined the Peace Block, a new political group which includes both men and women and which has formed the most substantial opposition to the Israeli government's policies of sealing off the Occupied Territories and engaging in war in Lebanon. These women are also trying to change the ways of organizing mixed groups of women and men.

Third, group work is beginning within the coalition on the topics of violence, militarism and feminism.

This period of decline in the scope of activity against the Occupation and of most women's withdrawal to the private sphere has constituted for many activists an embarrassment; nonetheless, it has also provoked renewed searching for a political platform which would be appropriate to these complex realities. Moreover, this period of transition has encouraged Israeli Jewish women to further develop their political feminism. In a joint seminar of Italian, Palestinian and Israeli women[7] which was held in Italy in

September 1992, discussions about nationalism, fundamentalism, and militarism enabled participants to enhance the depth of their political message; and at the yearly conference of Women and Peace in January 1993 these activists began to deal publicly with the connection between militarism and violence in Israel. For example, Dr Erella Shadmi said:

> occupation is, first of all, the symptom of a sick society; a society in which violence is legitimate and accepted from the social standpoint as a mechanism for personal expression and for solving conflict situations as well as for gaining rights and benefits. Occupation is possible in a society which justifies violence by creating its own myths and by the way it explains and defines its history.

At that same conference there was a call for the politicization of motherhood and for public political debate of participants' personal feelings and thoughts about their children's service in the army.

ISRAELI WOMEN AND NATIONAL SECURITY

The precedent for the political involvement of Israeli Jewish women in national security issues was set by Women Against the Invasion of Lebanon,[8] founded in response to the 1982 Israeli invasion. Then, for the first time in Israeli history, feminist women organized demonstrations against a state-sponsored war, challenging the legitimacy of war as a means for achieving political goals and linking the primacy of war to the secondary status of women in Israeli society. However, this radical critique did not gain mainstream currency among women in Israel at the time, and did not shape the agenda of women's peace groups until some five years later.

The political awakening of Jewish women in Israel has largely been an outcome of the *intifada*, which has posed a sobering challenge to the perception of the Israeli army as an army which serves in the role of defender only.[9] The *intifada* unveiled, first and foremost, the role of the Israeli army as the oppressor. As a result of the *intifada* it was no longer possible to believe, even as many still sought to do so, in the comfortable illusion of an 'enlightened occupation'. The fact that the *intifada* was a civil rebellion, and not a conventional war, throughout which the army faced an 'enemy' that included women and children, led many Israeli women into political opposition against the Occupation. In an interview with women of Women in Black from Kibbutz Megido, an Israeli woman named Shoshi says:

> At the beginning of the *intifada* . . . the atmosphere in the kibbutzim was similar to that which followed the Sabra and Shatila massacres [where hundreds of Palestinian refugees were massacred by the Lebanese Christian Falanges, with Israel's full knowledge]. There was an awakening and a fear for the shape of Israeli society. Kibbutz members said that something must be done. We have members of

German origin who were in great distress. They said it reminded them of fascism. People responded emotionally, they did not care under what flag they stood and with whom.[10]

Israeli women have gone into the streets to demonstrate when it has become clear to them that military action by the Israeli army has been aggressive and not defensive. This was the case during the war in Lebanon and, even more, during the *intifada*. The sight of soldiers chasing civilians, including women and children, created a crack in many Israeli women's perception of the army as moral and just. These women found themselves dealing with the question of what the men in their family – spouses, children, fathers, and others – were doing during their military service in the Occupied Territories, when their duties included the suppression of the uprising. Even though these women were not involved directly in the continuation of the occupation,[11] they heard disturbing reports about the brutality of the Israeli army, and through their own family members many saw themselves as partners and thus as responsible. For example, Daphna, a regular participant in the protest vigil of Women in Black whom I interviewed in August 1992, states the concern which she felt when her son was to be drafted during the *intifada*. 'Before his enlistment I was anxious. Those were the peak years of the *intifada*. I didn't want to see him either chasing or chased, nor either that they would throw stones at him or that he would arrest those who did.'

Some Israeli Jewish women went to the streets because they feared the effect that playing the role of the occupier and oppressor would have on their children and on Israeli society as a whole. Chava, another vigil participant whom I interviewed, was worried about the emotional price that her son and his friends might have to pay for their participation in the violent suppression of the Palestinian uprising. 'I can hear how his voice gradually changes into that monotonous army tone, utterly lacking any emotional color,' she said. Sarit Helman and Tamar Rapaport, two researchers from the Hebrew University, found in an as yet unpublished study that most Israeli women demonstrated publicly – some for the first time in their lives – in the name of universal values of justice which they saw threatened by the Occupation.

Organizers amongst women seeking peace recognized the importance of creating a political consciousness which would be acceptable to both sides, and of initiating discussions of political topics of existential importance, such as racism, prejudice and fear, which constituted obstacles to peace (Deutsch 1992). Due to the national conflict and the price it demands, especially from the Palestinians living under occupation, but also from Israeli society, the main topic on the public agenda was and still is a political solution acceptable to both peoples. Yet a women's protest and peace movement cannot be satisfied only with a political end to the Occupation within the framework of the patriarchal rules of the game. It has to create a political culture that challenges these rules and it has to unite the political and the personal on the basis of the life experiences and world visions of these women. In the meantime, a Jewish woman in Israel has to struggle with three main obstacles to raising her consciousness on the questions of the army and militarism:

1 A world-wide reality of nation states whose status rests on military strength and which settle disputes by means of war.
2 Daily life in a situation of national conflict, which creates an existential fear and increases Jewish historical anxiety.
3 The social expectations engendered by the traditional image of the protected woman (Hicks Steihm 1982) – and the accompanying lack of freedom to express judgments on the subjects of war and peace – and the obligation of contributing to the national effort and making maternal sacrifices.

In the consciousness of many Jews the establishment of the state of Israel was the political answer to the persecution and the helplessness of Jewish life in the Diaspora. Overcoming the sense of powerlessness and loss which stemmed from the Diaspora experience was accomplished through the building of a concrete national entity while ignoring the experience of feeling weak and hurt as well as the existing political reality in Palestine. The building of military strength – aside from its functional role in realpolitik – was used as a substitute for coming to terms collectively with the meaning of being persecuted. The army was, and to a great extent still is, one of Israel's main mechanisms for creating a common national identity. It has been central in creating and reinforcing the image of the new Israeli, the antithesis of the persecuted Diaspora Jew. The army became the basis of a positive male self-image and a source of national pride which emphasized independence, protection, and security. Belonging to the army has, as a result, symbolized belonging to the new society and state, a particularly significant issue in a society of immigrants and refugees.

Women are also affected by the desire to feel part of Israeli society by serving in the army. Both men and women are drafted into the Israeli army according to the Army Recruitment Law, and this policy, publicized by photographs of pretty, smiling women dressed in uniform and holding guns, has promoted the myth of the Israeli army as exemplary in its equality between the sexes. But, in reality, even though today there is an attempt to assign them professional roles, women mainly perform relatively minor service and administrative functions in the Israeli military – and therefore the meaning of the military service is different for them and for men.

The army is the main source of men's social mobility in the state of Israel. Building a military career leads to the acquisition of political and economic control, and active participation in combat creates a male elite in control of civilian life and provides citizen-soldiers with the credibility necessary for the expression of political criticism. This hierarchy is even present amongst the Israeli Jewish peace groups. For example, Peace Now, which was created by army officers who wrote in protest to then Prime Minister Menachem Begin in July 1978, did not allow one of the initiators, a woman officer, to sign their letter because she had not participated in an act of warfare.

Because they have marginal functions both in the army and in society's decision-making hierarchy, Israeli women cannot take an active part in setting policy and can at most serve as providers of traditional services within a patriarchal society. Awareness of this marginality has the potential to create the individual and collective feeling of freedom essential to shaping a feminist agenda; but the intricate social web within which they function as protected – as daughters, as wives, and as mothers of soldiers who protect them – frequently prevents women and the women's groups which they have established from achieving this potential (Enloe 1983). The process of collective feminist consciousness raising in support of peace remains fraught with personal conflicts for women. Dealing with feminist issues demands a daily examination of questions of personal identity and of activists' personal obligations, such as the obligation to educate their sons that serving in the army is an important national duty and personal ideal.

The army must be considered, on the one hand, as defined by the permanent reality and the fact that nation–states rest on political might; but, on the other hand, feminist criticism of militarism must create an ideological base for international political transformation. In the context of the Middle East conflict, the Israeli army is seen unequivocally by most Israelis as a framework which provides defense. Life in the shadow of this conflict and the violence which it entails evokes among Israelis considerable existential anxiety. In the same way that the individual overcomes anxiety by employing basic psychological defense mechanisms, so does the collective. The army has managed to convert itself into the national defense mechanism of a people who feel under siege – and so far, the peace work of the Israeli women's movement has been carried out within this basic framework of the political–cultural canon of patriarchal society without challenging either the basic tenets of the nation–state, which are based on national security, or their implications for relationships between the sexes. Even those Israeli women who seek peace in the full sense of the word still generally conform to these publicly legitimated perceptions of 'national security' without reaching any feminist consciousness about the link between those perceptions and the status of women. They are politically active because they feel a close link with their Jewish identity and with the Jewish collective. Yet they have little sense of linkage to a distinctly women's historical experience grounded in the arbitrary division between protectors and protected (Hicks Steihm 1982).

An example of this gap in Israeli women's feminist consciousness is provided by L. (who prefers to remain unidentified). On a purely ideological level L. is a deeply committed pacifist. She has worked intensively to promote peace between the Israeli and the Palestinian peoples, and at times she has had to pay a high personal price for this choice. Yet L. sees and accepts the army's function as ensuring Israeli national survival within the existing reality. 'Were it not for the army our very existence would be endangered. [This is] a real threat and not only a psychological one because of the Holocaust mentality.' None the less she also says: 'When I see what they do to the Arabs, my upheaval, my sorrow, my pain is terrible, it is awful! And I don't justify this in any way! . . . I refute this totally!' Similarly, during the conversation she says; 'on a purely ideological level I am all in favor of the elimination of all armies of the world' – but then she adds:

> [Despite] all the nice ideas of the world . . . I feel this in my gut! It is a point on which I stand firm. For me it refers to a loyalty to a general community . . . which is my people. It is a very primal energy . . . I also see in this a female aspect . . . the feeling that I want to protect: myself; my gut; and all that I have within, which is also my people. When it comes to the responsibility of a large body, of many people, I feel that I have no right to jeopardize the lives of others because of ideology, in this instance pacifism. I do not think along those lines at all in our case, although there may be other situations in the world where it is appropriate to begin with that. And I very much love Gandhi, really love him . . . It is a contradiction. I really believe that his path is the correct one, in its way and in all its ideology; however, when it comes to the gut and [when] one is talking of my reality in Israel, as a Jew, with all our history, in a totally emotional way . . . I am willing to have an army for survival but not to pay the price of the destruction.

L. helps us to see the dilemma posed by having to choose between, on the one hand, the ideology of a peace seeker and, on the other, the need to modify women's ideological positions with respect to a reality created from radically different values by men who are in a position of strength, be it in Israel or elsewhere. She insists upon differentiating between the defense of life, which is perceived as a value, and the complex meanings of the army, including its influence on the increase of violence in society. She strongly criticizes the violence of the occupation army; but she does not see this violence as an inevitable outcome of the existence of the army and of the male militaristic culture which creates it and is shaped by it.

Among the women in the women's peace groups, one can also hear voices which express a firmer stance against the existence of armies. Miriam is the daughter of American parents who emigrated to Palestine in 1946. Her father was killed by an Arab sniper in the 1948 war, while working on his kibbutz. About a year before her eldest son was to be drafted into the army (during the Lebanon war) Miriam felt engulfed by anxiety and began to wake up in the middle of the night with nightmares.

> [T]hat particular night I went to sleep as I normally do, but I woke up at 3:30 in the morning, in a cold sweat, my heart pounding fiercely. I will never forget it . . . . During the day I would function; however, every night at between 3:30 and 4:00 am I would wake up with those same symptoms of cold sweat, fast heart beats and most distressing thoughts.

Since that first night she has retained the habit of waking up in the very early hours of the morning.

> For me, the fact that my sons go to the army (my second one is due to be drafted in a few weeks) is very traumatic indeed, a true threat . . . I speak to many mothers of soldiers. They all share the same deep fear that something will happen to their sons while on military duty. I must admit that I do not have this kind of fear; I confess that the whole issue of the army, with all that it entails, is very difficult for me to deal with, and the fact that my sons have to take part in it makes it more so. Human beings created this system of the army for themselves. It must be a very primitive and sad society which needs an army to organize its life. If human beings created it, then they too can do away with it.

She immediately adds, struggling with the contradiction:

> [T]o totally do away with it, I believe, is impossible, but . . . the whole idea of sending our young boys to the army . . . their desire to become fighters, where does it come from? From home, from the street, from the school, from the general atmosphere. I feel that this is a grave mistake . . . my gut reaction is not to agree with the existence of the army, any army. I accept it in my mind only. I never internalized its necessity . . . . The army was created by human beings, so it can be dismantled by them.

Unlike Miriam, most Israeli women tend to fit their world view to *realpolitik* even when it is in conflict with their own ideology, which they perceive as irrelevant to existing realities. They carry within themselves the longings, the fears, the anxieties, the hurt, and the anger; and they ignore the threatening additional meanings of the army and the army's influence on the quality of their lives, on their families, and on society at large. The demand that their sons participate in the national defense effort in a world which is shaped by military strength is, in the eyes of most of these women, inevitable. Even if they identify with those who refuse to serve in the Occupied Territories, most Israeli mothers refrain from trying to exert influence on their sons, because they do not see themselves as the ones who have ultimately to decide and to pay the price in the sense of belonging to Israeli society and preserving professional possibilities for themselves. Daphna, for example, says about this issue:

> [R]efusing to serve [in the Occupied Territories] . . . I have no clear cut opinion which I can put forward. I respect those who refuse to serve and [I] also participate in their demonstrations. However I did not advise my son either one way or the other. If I cannot do it myself, then I have no right to recommend it to another.

Daphna was born in Israel, and was five years old during the 1948 war. While as a child she felt proud to live close to a military base, within four days after the end of the Six Day War she began to feel uncomfortable and fearful about Israel's victory and its potentially destructive impact on Israeli society. In those feelings and fears she felt isolated from her family and from Israeli society at large. Only with the outbreak of the *intifada* did she realize that her fears had become true. With respect to Israeli soldiers' refusal to serve in the Occupied Territories she says: 'They are heroes in the fact that they are ready to refuse and to pay the price.' The new heroes, at least within a limited circle in Israeli society, are no longer those who are willing to pay with their lives for the defense of their homeland but rather those who are willing to pay the price for not following the mainstream, because of a moral–political conflict. Quietly, almost surreptitiously, Daphna says that her son served for a few months in the Territories; but she immediately adds with satisfaction that since he speaks Arabic he would often serve as a negotiator between Israeli officers and the local Palestinians. He chose to serve in the artillery, and that eased his mother's conscience because that type of service can be defined as defense of borders and doesn't involve confrontation with the civilian population. Thus at the intellectual level Daphna demonstrates the beginning of consciousness about the violence and destructiveness inherent in the organization of the army. Yet while talking to her I found that she still wanted to believe in the fixed values of the moral army and the purity of arms, values that she received through her education, even though she says: 'Our army is not purer, not better and not more just than other armies.'

THE GULF WAR AND AWARENESS OF THE MARGINALITY OF WOMEN IN SOCIETY

Immediately after the end of the Gulf War, public meetings were held by establishment women's organizations in Israel to discuss the effects of the war on the position of Israeli women. There was talk of the fact that women had disappeared from the media during the war, and talk of the economic, industrial, and professional price paid for the war by women because of the closure of the education system. This was the first war in Israel in which Israeli men did not take an active role and could not prove their military prowess as protectors. This time, they sat together with the women, the elderly, and the children away from the front and experienced the feelings of powerlessness common among the 'protected' civilian population in times of war. The steady increase in the number of incidents of murder and battering of women by their spouses during the war was given public attention not only by women's groups but also by the media, the press, radio and television.

As feminist activists spoke out strongly, a public dialogue on violence against women began to emerge; yet no adequate public confrontation of the sources of social violence as a whole and in Israeli society in particular, and of the connections between militarism and violence, has yet taken place in Israel. In spite of the debate on domestic violence, very little has been heard in the popular arena against the use of war as a legitimate means of solving conflicts. These women did not speak of the need to change the existing rules of the game, but rather of the influence of existing rules on women and of ways of organizing to improve the status of women within existing frameworks. For example, with the increase in public awareness of the centrality of the army within Israeli society there came, in a great many of the meetings, agreement on the need for equal participation of women in the army as part of women's struggle for improving their status. Yet women's call for equal opportunities in the army also means accepting conformity with a patriarchal system of values which inherently fosters discrimination, deprivation, and violence, and thus using women as means to further the same sexual, emotional, and economic interests which marginalize them in the social decision-making process.

Only on a panel sponsored by Women and Peace in April 1991 in Tel-Aviv, at the end of the war, did Chava Lerman, a feminist peace activist, finally emphasize that 'there exists a diametrically opposed relationship between the level of militarism of the society and the equality of rights and the status of women it.' Since then, Israeli feminist peace activists have been trying to dedicate their time to linking the effects of the war to the status of women in Israel. Thus the *intifada* brought with it an increase in awareness concerning the role of the army in the oppression of the Palestinian people; and the Gulf War brought to the fore, in turn, the centrality of the army in the Israeli society as an instrument for marginalizing women.

THE CHALLENGE: THE CREATION OF A FEMINIST POLITICAL CULTURE OF PEACE

Since the army has retained such a special place in Israeli Jewish society, the women's peace movement has not yet been able to create a political alternative to existing reality governed by patriarchal values like war, occupation, oppression, exploitation, violence, and rape. Women need to use the recent increase in awareness of violence in Israel to deal with the centrality of the army in Israeli society, and to identify publicly and decisively the link between militarism and the status of women in Israel.

The Israeli military occupation of the West Bank and Gaza Strip has played a key role in raising the political consciousness of Israeli Jewish women and providing them with an arena within which they have been able to develop important political skills. Now the women's peace movement within Israel must find effective ways to oppose the Occupation while ceasing to view the Occupation as the root of all evil and seeing it instead as a symptom of a militaristic, patriarchal society.

Today, after more than six years of activity, Israeli feminist peace activists stand at a crossroads. The women's peace movement in Israel now faces the challenge of establishing an ambitious political stance which will connect the status of women to militarism in Israeli society. The link between the personal and the political is essential to free women from the bonds of the internalized values of militarism: first and foremost, Israeli women need to free themselves from the role of 'protected women' who cooperate with men who act as if they protect them. Finally, in the contemporary Middle East, breaking patriarchal bonds is especially difficult. The reality of the Occupation and the fundamental national conflict focuses Jewish and Palestinian women's attention on aspects of national identity. In the existing tension between national and gender identities, national identity tends to assume the greater importance. Although it is well known that during national struggles women's movements flourish, it is equally well known that these movements are usually short lived. The importance of national women's movements tends to erode over time, especially when women's contributions to the national struggle are no longer needed, as in the case of Algeria; when women's activism becomes a threat to the male leadership, as in the Palestinian case; or when women's issues become marginalized.

Today, when there is hope for an apolitical solution, in the wake of the joint Israeli–PLO recognition, the women's peace movement still has a dual role. First, women must make sure that painful political topics which have existential implications for the relationship between Palestinians and Jews, such as the fate of Jerusalem and Palestinians' right of return, remain on the agenda of public debate. Only a public confrontation of the meaning of these differences between the two societies will bring about reconciliation between the two peoples. Second, in order to create a feminist political peace culture, women must create political and social transformation, based on the unique yet diverse life experience of women and on the inner, whispered world of women which has not yet achieved public expression. If women indeed wish to achieve meaningful change, only joint work among women of all states in the Middle East and in the world can mount a credible challenge to existing patriarchal priorities.

NOTES

1 This chapter is based on a lecture entitled 'Conflict, War and Militarism' which I delivered at a joint feminist seminar of Italian, Israeli, and Palestinian women, held in Italy in September 1992 at the initiative of the Center for Documentation of the Women of Bologna in cooperation with the Women's House of Torino.
2 The women's peace movement in Israel includes Jewish women and Palestinian women from within the pre-1967 borders. In this chapter, reference to Israeli women includes both Jewish women and Palestinian women who reside in Israel. It is worth noting that large numbers of Palestinian women in Israel do not define themselves as Israeli, but rather as Palestinian with Israeli citizenship. In this chapter references to Palestinian women are intended to mean Palestinian women from the Occupied Territories.
3 Casa delle Donne is the Women's House in Torino and Centro di Documentazione is the Center for Documentation in Bologna.
4 In an attempt to pierce the apathy and indifference of the average Israeli a women's group of veteran feminists screened slides showing the brutal reality of the Occupation weekly in a central street of Tel Aviv (Ostrowitz 1990).
5 The coalition includes women from Women in Black; Women for Women Political Prisoners; *Shani* – Women against Occupation; *Tandi* – Movement of Democratic Women; the Israeli branch of WILPF; Women for Co-existence (*NLD*); and Woman to Woman.
6 Within the general extra-parliamentary groups such as Peace Now, women play important roles in the organizational work but rarely speak in public.
7 The seminar was initiated by *Casa delle Donne*, Torino, and *Centro di Documentazione Donne*, Bologna and it was attended also by British and Algerian feminists active against fundamentalism.
8 Another known group from the same period is Mothers against Silence. Although this was originally a joint male and female movement, Parents against Silence, the men gradually dropped out.
9 The concept of defense is also expressed in the army's name: the Israel Defense Forces.
10 See Lahav Hadas, 'Women in Black': September 1990: 26–27.
11 It is important to point out that, especially during regular army duty, Israeli women serve in the Occupied Territories and thus take an active part in the continuation of the Occupation.

BIBLIOGRAPHY

Brook-Utne, B. (1985) *Educating for Peace*, New York: Pergamon Press.
Chazan, N. (1991) 'Israeli women and peace activism', in B. Swirski and M. Safir (eds) *Calling the Equality Bluff: Women in Israel*, New York: Pergamon Press, 152–162.
Deutsch, Y. (1992) 'Israeli women: from protest to a culture of peace', in D. Hurwitz (ed.) *Walking the Red Line*, Philadelphia: New Societies Publishers.
Enloe, C. (1983) *Does Khaki Become You? The Militarization of Women's Lives*, London: Pluto Press.
Espanioly, N. and Sachs, D. (1991) 'Peace process: Israeli and Palestinian women', *Bridges*, 112–119.
Hicks Steihm, J. (1982) 'The protected, the protector, the defender', *Women's Studies International Forum*, 5, 3/4: 367–376.

Lahav, H. (1990) 'Women in Black – Megido Junction: a collective picture', *Challenge*, 4, 26–27.
Ostrowitz, R. (1990) 'Dangerous women: the Israeli women's peace movement', in R. Falber, I. Klepfisz and D. Nevel (eds) *Jewish Women's Call for Peace*, Ithaca: Firebrand Books.
Sharoni, S. (1992) 'Every woman is an occupied territory: the politics of militarism and sexism and the Israeli–Palestinian conflict', *Journal of Gender Studies*, 1, 4, 447–462.

6

PALESTINIAN WOMEN IN ISRAEL
Identity in light of the Occupation
Nabila Espanioly

Very little research has been done on the Palestinian woman in general or on, more particularly, the Palestinian woman in Israel. The role and status of women within Palestinian society in Israel are to a great extent determined by political, economic, social and ideological forces. The woman in Palestinian society cannot be separated from her place within the family, for the family remains a major force in shaping a woman's identity, determining whether or not she can marry as well as her relationship to people inside and outside her family. In many Palestinian families, moreover, depending on class, social and political status, women move upon marriage from the custody of their fathers and brothers to the custody of their husbands and their husbands' families. At the same time, because Palestinian society in general, and Palestinian society in Israel in particular, has undergone dramatic social, political and economic changes, women's position within the family and within society has changed as well. Like Arab women elsewhere, Palestinian women in Israel have moved in recent years away from traditionally passive and marginal roles into more active ones (Ginat 1982; Touma 1981; Nath 1978; Smock and Youssef 1977; Prothro and Diab 1974), both with respect to their Palestinian identity and the Israeli occupation of the West Bank and Gaza Strip, and with respect to traditional structures within their own society.

The first public participation of Palestinian women alongside Palestinian men in the national fight, in civil and human rights, was recorded in 1884 (Warnock 1990). More than 100 years have passed since this first involvement, and Palestinian women continue to play an integral part in the struggle of the Palestinian people for self-determination.

Palestinian women in Israel have experienced substantial changes, some of them common to the changes all Palestinian women have experienced and some of them unique to the Palestinians who live in the State of Israel. The changes that started during the Ottoman rule and which have continued under Israeli rule have left their mark on the Palestinian people, including the Palestinians who live in Israel, and an understanding of this historical background is essential to an understanding of the changes which Palestinian society in Israel is now experiencing.

In this chapter we will examine the Palestinian people during different historical periods without considering separately their different constituent groups. For only after 1948 were the Palestinian people divided into the following four parts: Palestinians who remained living in their homeland (which became the state of Israel); Palestinians who lived under Egyptian rule (in Gaza); Palestinians who lived under the Jordanian rule (in the West Bank); and Palestinians who live in the Diaspora. Here, we will be examining primarily the Palestinians who live in Israel and their different struggles. We will look specifically at the women, their political experience and their changing gender, national and class identities which are constantly developing and influencing their position and ability to organize within their society.

The Palestinians in Israel, including the women, underwent a process of politicization as a result of the war of 1948 and the Occupation of 1967. The latter in particular made them confront major questions regarding their identities – and in their attempt to answer these questions and to respond to the new needs which resulted from the social and political changes taking place, Palestinian women in Israel developed new ways to organize. The Occupation itself and the *intifada* are but two important stages in this process of empowerment. To understand this process it is important first to identify and then to examine the social precedents and patterns which have evolved historically.

THE ROLES OF PALESTINIAN WOMEN: HISTORICAL PERSPECTIVE

Ottoman period (early sixteenth century–1918)

During the Ottoman period, Palestinian society was a largely peasant agricultural society. Working the land was the basic means of subsistence and survival for the majority of the population. The landowners held the ultimate power over the land through different Ottoman laws (Abdo 1987). In working the land the peasants, both men and women, provided an income for the state and the landowning class, as well as for the whole *hamoula* (extended family).

The *hamoula* was the main social, economic, and political unit where the *hamoula*'s chief was the main decision maker in all aspects of life. The structure of the family and society was very hierarchical, and the dominating norms were patriarchal. Although the women worked the land together with the men of their *hamoula* (normally one village consisted of one or two *hamoulas*), unlike men they did not have any social or political power. In fact women adapted the forms, norms and attitudes of the patriarchal system into their private life. The wife of the head of the family, for example, assumed command over all other women in the family, using the same hierarchical order employed by men.

Inequality based on gender differences was more prevalent among the upper, landowning classes within which women were largely segregated and confined to the domestic sphere. In this period, women of the landowning classes adopted the Ottoman habit of veiling their faces in an attempt to differentiate themselves from the peasant women (Abdo 1987). Peasant women, who worked the land together with men, enjoyed some social advantages over urban women; yet they nonetheless were situated at the bottom of the social–political structure of the Arab society of Palestine.

British mandate as transformation period (1918–1947)

At the end of the Ottoman period and the beginning of the period of the British mandate over Palestine, social, political, and economic developments led to significant structural changes in the family, resulting in a crisis for the peasant class and the beginning of a transformation in women's roles in Palestinian society. The economic–political crisis sparked by transition motivated landowners to sell their land to investors, thus leaving the peasants' families without basic means of subsistence. Thousands of peasants found themselves landless and were forced to leave their land and villages and to seek work and shelter in the cities. New structures of landownership and methods of farming throughout British-administered Palestine had the effect of strengthening the merchant class. This shift of power to the capitalist class was further accompanied, and influenced to some degree, by an influx of European Jewish settlers who brought with them industrialized resources and capitalist aspirations.

These fundamental economic changes initiated a process of proletarianization among the Palestinian peasants – and, ultimately, caused serious disturbances in the traditional structure of the extended family. Many landless peasants' families increased their dependence on wage labor, which resulted, in turn, in a decline of traditional dependence relations within the family, and in migration to towns and depopulation of the rural areas.

These social and political changes, and related fear for their country's future, motivated urban Palestinian women, especially of upper and middle classes, to take some form of action (Rishmawi 1988). For the first time, Palestinian women began to engage in social activism, organizing charitable societies in the major cities of Haifa, Akka, Jaffa, Nablus and Jerusalem (Encyclopedia Palestina 1990), mainly during the years 1904–1916. After years of activism at the local level they gathered on October 26, 1929 in Jerusalem for the first Palestinian women's conference (Giacaman and Odeh 1988; Rishmawi 1988; Fawzia 1984). Generally speaking, those women who were motivated to come out on to the front line and to participate in the national struggle were wives or relatives of men who were politically involved (Abdo 1987; Rishmawi 1988).

Palestinian women were also affected by the education boom in the Arab world during the late nineteenth century which resulted in the spread of schools, especially missionary institutions. In addition, other educational institutions for girls were founded later in this period: in 1924, for example, Nabiha Nasser founded the Birzeit school, later to become Birzeit College and in 1976 Birzeit University (Rishmawi 1988). Yet even though demand for girls' education was increasing as Palestinians began to realize the important role education would play in helping them survive in a fast-changing and threatening world, the British mandate government did not do a great deal to improve the situation. By the end of the British mandate, one-third of Palestinian school-age children were in school, of whom one-fifth were girls (Warnock 1990).

The women of the poor classes were engaged in different forms of militant participation especially during the Palestinian revolution of 1936–1937, but more on a individual rather than on an organized basis (Abdo 1987).

Israeli military rule over the Palestinian community in Israel – a process of underdevelopment

The increasing participation of Palestinian women in public, social and political activities stopped short as a result of the 1948 war, especially among those women who remained within the new state of Israel. The war destroyed the social, political, and economic infrastructure of Palestinian society. More than 480 (out of 573) Palestinian villages were totally destroyed. Of the Palestinian population 75 per cent (750,000) became refugees in neighboring Arab countries. They were forced to leave by the Jewish forces (which later became the Israeli army), although many hoped to come back when the war was over. Only 150,000 Palestinians were able to stay within the new state of Israel. Of those who stayed, 40,000 of them found themselves refugees in their own land. Altogether, the Palestinians who remained in Israel found themselves in a shattered society, whose internal economic and political institutions and organizations had collapsed and whose cultural traditions were being threatened and challenged. Faced with this upheaval and with the problem of surviving under these new circumstances, the Palestinians who remained in Israel embarked on a process of great change.

Coupled with the Israeli government's policy of massive land expropriation (Jirys 1976), the effects of the war made it impossible to reclaim agriculture as the mainstay of Palestinian life. At this stage Palestinians faced increased proletarianization and impoverishment, and experienced what Abdo calls underdevelopment and paralysis (1987). This process included the destruction of the social and political infrastructure and the relatively autonomous basis of Palestinian economic life, and it transformed

Palestinians into a minority totally dependent on the Jewish-dominated economy. Moreover, Israel introduced a sophisticated system of hegemony which included a policy of control and manipulation aimed at undermining the integrity of Palestinian national existence within the newly established state (Lustick 1980).

Palestinians in Israel were isolated from other Arabs and segregated from Jews through military laws (Peled 1992). These laws controlled the daily life of Palestinians. According to these laws Palestinians in Israel could not, for example, leave their villages without a permit from the military authorities. In addition, because their land had been confiscated by Israeli authorities, Palestinians in Israel were forced to search for work outside their villages, and in order to do so they were forced to obtain military permits (Jirys 1976). Moreover, while the Israeli government granted citizenship to Palestinians in Israel, it did not recognize them as a national group; instead, it called them 'Minorities', 'Israeli Arabs', 'Non-Jews' but never 'Palestinians'.[1]

Under such traumatic conditions the feelings of insecurity, especially among men, were frequently overwhelming. Their continuous absence from home, their subordination to Israeli Jewish institutions, and their exposure to the behaviors of Jewish westernized women all intensified the threat both to male identity and to male status (Mar'i and Mar'i 1985). Having lost control over their land, status, present and future, the Palestinian man was left in control of only one domain: his family, his wife, and his children; as Mar'i and Mar'i stated, 'One can relate much of men's need to control women to men's sense of insecurity' (Mar'i and Mar'i 1985). This response was reflected in a renewed emphasis on questions of morality. In particular, the concept of *Ard* (Honor) acquired new importance and meaning in light of men's fears, their sense of threat and powerlessness.

As both the present and future became uncertain, and as Israeli authorities stepped in to control the lives of the Palestinians within the new state's borders, the heritage of the past became the most salient source from which Palestinians in Israel could derive pleasure as a community and on which they could depend for protection and preservation of their identity. Here again women have had to deal not only with the outcomes of land confiscation, military laws and other externally imposed objective hardships but also with the internal, subjective hardships reflected by centuries-old patriarchal tradition, which now gained nationally sanctioned importance (Abdo 1987).

Since Palestinian women in Israel were forced to stay at home because of military orders and because of restrictions placed on them by the men in their families, they were no longer able to support their families as producers, and they lost much of their previous status. Relegated to the private sphere and to domestic roles, Palestinian women in Israel assumed the role of preservers of culture. As such, they were expected to maintain the continuity of Palestinian cultural values, and thus to pass on traditions and values which reproduced their own subordinate conditions. Shaloufeh (1991) analyzes this situation according to the theory of Kreiter and Kreiter (1978), and shows that the Palestinian woman has been assigned conflicting tasks. On one hand, she is party to a belief-system which is consistent with a positive image of the Arab woman as active in the preservation of cultural, religious and national continuity; on the other hand, she is asked to accept and internalize a belief-system that defines her own status as inferior. In other words, a considerable portion of the cultural value system which Palestinian women have been assigned responsibility for upholding consists of the same values which have discriminated against them and deprived them of status for many generations. At the same time, one might argue that there is no real contradiction between the two tasks, because both involve adapting the role of culture preserver in the domestic sphere, including those parts of the culture which reproduce the condition of women's own subordination. Palestinian women act within the patriarchal system of their society as a major control mechanism, working alongside other agents of other control, like family and religious institutions, to preserve the patriarchal structure and thus the status quo of society in general.

While some positive changes took place during this period – the Israeli Compulsory Education Law increased the rate of school attendance in the Palestinian community in Israel by both boys and girls (Yisraeli 1980), and compulsory education increased the demand for Palestinian teachers and created opportunities for women to work – on the whole, women's political and social status did not improve. On the contrary, no Palestinian women's organization was established during this period except for the Democratic Women's Movement, which was a mixed Arab and Jewish women's organization. And while the economic recession (1965–1967) produced economic pressures which pushed Arab women back into the work force, these women served mostly as unskilled labor.

ISRAELI PALESTINIAN WOMEN AFTER THE OCCUPATION

1967–1975: renewed shock and adjustment

The Israeli occupation of the West Bank and the Gaza Strip and of the Syrian Golan Heights after the 1967 war came as a terrible shock to many Palestinians, who had lived for years under the illusion that the Arab countries would eventually liberate them. The 1967 war destroyed this illusion, as well as that of a united Arab front bound by Egypt's Gamal Abdul-Nasser.

After 1967 a new economic–political situation developed as demand for Israeli products and employment opportunities for women, as well as for men, increased. This process led to increased awareness among Palestinian women in Israel of discrimination based on their nationality and class – which, in turn, sparked increased political action on their part. These changes were manifested in the growing participation of Palestinian women in Israel in demonstrations and other public activities as they entered the Israeli labor market and the Israeli educational system and as they joined with men in intensifying demands for full national and civil rights and for recognition as a national minority (Mar'i and Mar'i 1985).

Most important, the Israeli occupation of 1967 re-united Palestinians in Israel with the most important segments of the Palestinian people, dramatically encouraging the process of re-Palestinianization. Meetings between Palestinians from Israel and from the Occupied Territories marked the political renewal of Palestinians in Israel, enabling them to openly identify as Palestinians and at the same time to begin demanding rights as a national minority within Israel, a process which continues today.

These meetings between the Palestinians who live in Israel and their brothers and sisters, who had lived under Jordanian rule and who became occupied in 1967 by the Israelis, opened new doors for the Palestinian community in Israel. If before 1967 they lived under Israeli military rule (from 1948 to 1966) and had become totally disconnected from the Palestinian people and from the Arab nation, the Occupation of 1967 enabled Palestinians living in Israel for the first time to re-connect with themselves, their people, and their culture. As a result of the occupation of the West Bank, the Gaza Strip and the Golan Heights, new literature and news from the Arab world made their way into Israel, despite Israeli censorship. Up to this point they were simply the Palestinians who had remained in their land; but the Occupation of 1967 created the need for self-determination *vis-à-vis* both the rest of the Palestinian people and the Jewish citizens of Israel. As a result of the 1967 war, Palestinians living in Israel started to ask nationally motivated questions such as 'who are you?', 'what is the relationship between us and the Palestinians under occupation?' and 'what is the connection between us and the state of Israel?'. The answers to these questions increased the sense of national belonging among the Palestinians living in Israel, as well as the need for national rights as citizens and as a recognized minority within the state of Israel. This renewed sense of national identity was reflected in widespread public political activity in which women were important participants.

In addition to embarking on this process of re-Palestinianization, the Palestinians living in Israel have as a result of the Occupation become more politicized and a great deal more realistic. The 1967 war made Palestinians living inside Israel realize that help would not be forthcoming from the outside, and that they would have to find resources within themselves. It was then that they began to create their own political program. The Palestinian women in Israel who were influenced by this new reality organized first at the local level, in women's organizations which continued the century-old tradition of charitable organizations representing the bourgeois approach of offering aid to others. At the same time, Palestinian women in Israel also became involved at the national level as they joined national political organizations. They created new, highly political women's organizations which were affiliated with existing political parties or groups. Women's participation in different aspects of life increased; for example, from 1969 to 1972 the number of female Arab students enrolled in Israeli universities more than doubled – from 141 to 305 – while enrollment of male Arab students increased by only 25 per cent, from 450 to 565 (Mar'i and Mar'i 1985). For the first time women became involved in Arab Student Unions, known for their high levels of political activism, and women voted for and were elected representatives and executives in these unions.

Despite the improvement in women's position during these years and their greater participation in the labor market, the educational system, and social and political organizations, the standing of Palestinian women in Israel remained low. According to my feminist arithmetic, Palestinian women in Israel rank tenth in status among citizens in Israel society, after (1) Ashkenazi men, (2) Ashkenazi women, (3) Oriental men, (4) Oriental women, (5) Russian men, (6) Russian women, (7) Ethiopian men, (8) Ethiopian women, (9) Palestinian men. Palestinian women in Israel suffer from discrimination not only because of their gender, but also because of their national and class identity.

1975–1987: Israeli Palestinian women taking their case in their own hands

Palestinians within Israel continue to be discriminated against within Israeli society (Haider 1990, 1991, 1993; Zureik 1988), where they have the lowest class status (Ben-Porath 1989). Many years of political activism and of changes within Palestinian society which have accompanied the Israeli military rule and the Occupation have raised the consciousness of Palestinian women in Israel with respect to their oppression as part of a national minority and as part of their specific class. As a result, they have become more active participants not only in the sphere of family decision making, but also in the spheres of public life which have been traditionally prohibited to them (Touma 1981). This is a period in which Palestinian women in Israel began adopting a new approach: taking their fate in their own hands and acting to change it.

Palestinian women in Israel who have managed to acquire a higher education (Al-Haj 1988) have had to overcome many social obstacles, although recently there have been signs within Palestinian society in Israel of greater openness to women's working outside the home, especially in white collar occupations. In the 1986–1987 academic year, women constituted approximately 48 per cent of all the Palestinian students studying at Haifa

University. Women constituted 21.6 per cent of all Palestinians who have earned a first degree in Israeli universities, and 32.5 per cent of all Arab academics in Israel (Al-Haj 1988). Meeting both with Palestinian women of the Occupied Territories and with Jewish women in Israel has increased the feeling of oppression among Palestinian women in Israel. In comparison to the awareness of Palestinian women in the Occupied Territories the achievements of Palestinian women in Israel seemed weak. For example, despite the increase in education among Palestinian women in Israel because of compulsory education in Israel, the success Palestinian women had at Haifa University remained an exception: in 1986 the participation level of Palestinian female students in higher education had reached only 20 per cent among the Palestinian students in Israel (Central Bureau of Statistics 1989); while it had reached 40 per cent among the Palestinian students in the Occupied Territories (Central Bureau of Statistics 1982, 1985). This disparity, together with the fact that they are the lowest paid members of the Israeli labor force[2] (Central Bureau of Statistics 1992), raised their awareness of oppression, and this very consciousness became the prime factor in organizing efforts among Palestinian women in Israel.

During this period Palestinians in Israel organized nationally within Israel, and escalated their demands for rights and for peace. They created new committees, which were male originated and dominated, such as The Follow-up Committee for Arab Issues (which included all Arab mayors and representatives of Palestinians living in Israel), The Follow-up Committees for Arab Education and Health Issues, and the Committee for the Protection of Arab Land, to mention a few. All these committees reflect the politicization and re-Palestinianization of Palestinians living in Israel.

Women in their own way also organized different committees to support Palestinian women in Israel and to develop local services, especially in areas which were important in the light of their new political reality and which were neglected by Israeli authorities, like the educational framework for preschool children and for women, and committees whose goals were to support women. Many of these committees, like the Nazareth Nurseries Institute, or the Akka Women's Association–Pedagogical Center, have become important associations supporting Palestinian women in Israel. These organizations have played an important role in creating infrastructure for professional support for women and for empowering women.

In order to change their social and political situation Palestinian women in Israel shifted their strategy from demonstrations, which were typical of Palestinian activity prior to this period, to positive action. They took matters in their own hands as a way to achieve both local changes, within their own communities, and state-wide change which would lead to recognition of political rights for Palestinian people within their independent state.

Palestinian women in Israel, like the rest of the Palestinian community in Israel, faced a fundamental conflict between their civil identity as citizens of the state of Israel and their national identity as part of the Palestinian people and, as such, subject to oppression and suppression by the state of which they were citizens. To date, this conflict remains a crucial one for many Palestinians in Israel, and Palestinians in Israel differ in the ways they attempt to resolve it. For those who see themselves as sharing the same national identity and the same fate as the Palestinians of the Occupied Territories, the solution is to eliminate any differences between the two populations by establishing a state over all of Palestine. For those who see themselves as full citizens of Israel, the solution is to eliminate differences between themselves and the Jewish population in Israel, and to strive for full equality. The largest group of Palestinians in Israel attempts to solve the dilemma of identity by asserting both that they are an integral part of the Palestinian people and that they are unique because of the historical circumstances of remaining on their lands and becoming Israeli citizens. Their solution distinguishes between their civil identity as Israeli citizens and their national identity as Palestinians. As Israeli citizens they demand civil rights and acceptance as equal citizens, while as Palestinians they demand national rights as a national minority within Israel.

Each of the solutions to this conflict of identity is reflected in the different activities and organizations on the part of Palestinians in Israel. The last-mentioned solution enables many Palestinians in Israel to be critical of the country of which they are citizens and of its policies toward the people of which they are part. But at the same time this solution of remaining both Palestinian and Israeli citizens also enables mutual activities between Palestinians and Jews who are critical of the Israeli government policies so that they can work together towards political change for the Palestinian people.

The situation and the political awakening of Palestinian women in the Occupied Territories also contributed to the awakening of Palestinian women in Israel. Palestinians in Israel have intensified their efforts on behalf of and in solidarity with their brothers and sisters in the Occupied Territories. Palestinian women in Israel have been especially active in solidarity and protest activities (Espanioly 1991), including joint activities with the peace camp in Israel and with Israeli women's peace groups such as Women in Black, Women and Peace, and Women for Political Prisoners.

The *intifada*: 1987 until today

Palestinian women in Israel have watched carefully the changes among their sisters in the Occupied Territories and learned from their experiences. They have seen, for example, that Palestinian women's national awareness of discrimination has intensified under the Occupation. These conditions were precisely the factors against which Palestinian people in the Occupied Territories organized their protest and uprising. Palestinian women in Israel who are engaged in peace activities also tend to be motivated primarily by national considerations rather than by considerations of gender. When a Palestinian woman in Israel sees a Palestinian woman in the Occupied Territories facing a soldier, she tends to identify with her and to feel proud of such a woman just as she would feel proud of a Palestinian man challenging the Israeli authorities. Women in the Occupied West Bank and Gaza have actively participated in the *intifada*; and in the beginning stages of the *intifada* their participation was not only socially accepted but even highly appreciated. But in the later phases of the *intifada* the women who were active began to see the obstacles placed in front of their further achievement of positions of leadership and more militant, non-traditional roles. Palestinian women began to feel the restrictions placed upon them by religious fundamentalists and by conservative traditionalists, as well as by some of the 'revolutionaries' who were demanding that women sacrifice fighting back on their own behalf until after the revolution. Very few Palestinian women in Israel yet recognize the contradictions of the 'revolutionary' man who speaks day and night about equality, but then goes home to his wife, mother or sister and begins to act like a 'sheik' who needs to be waited on and made to feel that he is the boss.

This situation could continue indefinitely, because awareness of one's role as a woman and of the oppression one suffers at the hands of one's own man and one's own society is frequently more painful than awareness of the oppression suffered in common with one's people at the hands of the 'enemy'. Within Palestinian society in Israel women's identity is formed, that is, misinformed, by stereotypes and rigid norms through the patriarchal control systems of family, religious institutions and social institutions, and thus women themselves have frequently internalized a sense of their own inferiority. Yet Palestinian women's national identity in Israel has actually been nurtured in important ways by discrimination. The Israeli military occupation has intensified among Palestinian women and within Palestinian society as a whole a sense of national awareness and, to some degree, of class awareness.

When sexual harassment began to become an integral feature of Palestinian women's experience at the hands of the Israeli security forces (Senker 1989; Nevo, 1989), Palestinian women's organizations began to deal with these issues and to provide support systems to victims. These victims were not only suffering from their horrible experience at the hands of the Israelis but also from the reaction of their immediate society, which responded by blaming them. Whereas in some earlier cases young men had offered to marry victims – in an effort to support them and encourage them so that other women could continue the struggle – increasingly Palestinian families responded to Israeli sexual harassment by imposing greater control over women, by forbidding them from leaving home, by curbing their studies, and even by violence.

Sexual oppression within the Palestinian society as a whole remains present, whatever the woman's marital status; sexual relations are considered acceptable only within marriage; and women are supposed to remain virgins until they marry (Fawzia 1984). After marriage, the woman is expected to be sexually available to her husband at all times. She has no right to initiate or participate actively in lovemaking herself, and it is unlikely that she ever has an orgasm. In particular, speaking about sexuality and sexual behavior is strongly taboo within the Palestinian society. Violence and sexual assault within the family or outside the family bring shame on the victims, who tend to blame themselves and to keep the assaults, especially the sexual assaults, secret. But when sexual assault became a weapon used by Israelis against Palestinian women, particularly in the *intifada*, it became easier for Palestinian women to speak out about their sexual treatment and concerns. Although this process of speaking out remains as yet limited to individual victims and their support networks, it is likely eventually to have greater, more visible effects.

Many Palestinian women from Israel have become involved with Jewish activists in Israeli women's peace organizations and in feminist analyses of the effects of war and militarization on Israeli society [see, for example, the chapters by Sharoni and Deutsch in this book, Chapters 7 and 5 respectively]. The Palestinians in Israel who have worked with these groups have adopted an increasingly feminist perspective on the Occupation and its effects. The Palestinian women in Israel who have begun to criticize Israeli society, militarism, sexism and violence have begun, in turn, to see these phenomena more clearly in their own society and to speak out against discrimination and violence, especially domestic violence, and against sexual assaults of all kinds.

Perhaps most important, the Occupation and the *intifada* have simultaneously reinforced the process of re-Palestinianization among Palestinians in Israel, and at the same time, reinforced awareness of the difference in status between the Palestinians who are citizens of Israel and the Palestinians who are under the military occupation. Increasingly Palestinians in Israel are demanding not only their rights as a national minority within the state of Israel but also the right of their people to national self-determination.

CONCLUSION

While the direct effects of the Israeli Military Occupation and the Palestinian *intifada* on Palestinian women in Israel still need to be more thoroughly studied, some likely implications already seem evident.

First, coming at the end of over fifty years of shifting geopolitical relationships, the Occupation and the *intifada* have reoriented Palestinians in Israel in national solidarity with Palestinians in the West Bank and the Gaza Strip. Second, while this dramatic process of re-Palestinianization has been for some Palestinian women in Israel a source of frustration – 'We felt helpless because we could not support our sisters', one responded in a survey which I conducted in Nazareth (an Arab town in northern Israel) – the *intifada* especially has also very clearly politicized many others: 'It caused more involvement in the political arena'; 'it had the effect of empowering and giving [us] strength for social and political activities.'

Finally, contact with Palestinian activists from the Occupied Territories, with Israeli Jewish feminists, and with repressive Israeli political tactics has awakened among Palestinian women in Israel a consciousness of gender inequities within their own society. Again these responses, including responses to escalating conservatism and Islamic fundamentalism within Palestinian society, have crystallized around the *intifada*: 'It provides us with knowledge of our rights'; it brought 'liberation from the old social norms,' as some Nazareth Palestinian women, for example, stated. New awareness of women's activism and new organizations, including rape crisis centers and campaigns against violence against women, have been created, mainly by young educated Palestinian women in Israel who are now working to create a feminist approach to discrimination within Palestinian society, in the home and at the work place. At both the local and the national levels these women's organization are still struggling toward self-definition; while, for example, some state clearly that they are feminist organizations, their forms and structure still tend to replicate the hierarchical systems in which these women were educated and against which they are attempting to work. Moreover, the level of feminist consciousness among Palestinian women in Israel remains diverse, as differences, as well as apparent contradictions, among organizations reflect. Yet while these factors make it difficult to conclusively assess the direct impact of the Israeli military occupation on Palestinian women in Israel, it seems clear nonetheless that the Occupation and the *intifada* which has followed it have helped in important ways to politicize, re-Palestinianize, and raise the feminist consciousness of Palestinian women in Israel, whose lives will be changed in complex ways by this process.

NOTES

1 After 1967, and because of contacts both with the other parts of the Palestinian community and with the Israeli Jews from whom they had been disconnected until 1966 because of the military rule, their identity was sharpened and became the subject of a public debate.
2 Palestinian women in Israel earn only 60 per cent of what Jewish women in Israel earn. (And it is important to remember as well that Jewish women earn only about 70 per cent of Jewish men's incomes.)

BIBLIOGRAPHY

Abdo, N. (1987) *Family, Women and Social Change in the Middle East: The Palestinian Case*, Toronto: Canadian Scholars' Press.

Al Haj, M. (ed.) (1988) 'Problems of employment for Arab academics in Israel', *Middle East Studies*, 8, The Jewish–Arab Center University of Haifa.

Al-Khahli, G. (1977) *Palestinian Women and the Revolution*, Akka: Dar Al Aswar Publications (in Arabic).

Ben Porath, A. (1989) *Divided We Stand: Class Structure in Israel from 1948 to the 1980s*, Westport, CT: Greenwood Press.

Central Bureau of Statistics (1982, 1985, 1992) *The Statistical Yearbook for Israel*, Jerusalem.

Encyclopedia Palestina (1990) *Special Studies: Volumes III and IV*, Beirut: Encyclopedia Palestina Corporation.

Espanioly, N. (1991) 'Palestinian women in Israel respond to the Intifada', in B. Swirski and M. Safir (eds) *Calling the Equality Bluff: Women in Israel*, New York: Pergamon Press, 147–151.

Fawzia, F. (1984) 'Palestine', in R. Morgan (ed.) *Sisterhood is Global: The International Women's Movement Anthology*, Garden City, NJ: Anchor Books, 536–539.

Giacaman, R. and Odeh, M. (1988) 'Palestinian women's movement in the Israeli occupied West Bank and Gaza Strip', in N. Toubia (ed.) *Women of the Arab World*, London and New Jersey: Zed Books, 57–68.

Ginat, J. (1982) *Women in Muslim Rural Society: Status and Role in Family and Community*, New Brunswick, NJ: Rutgers University, Transaction Books.

Haider, A. (1990) 'The Arab population in Israeli economy', in D. Kretzmer (ed.) *The Legal Status of the Arabs in Israel*, Boulder, CO: Westview Press.

—— (1991) *Social Welfare Services for Israel's Arab Population*, Boulder, CO: Westview Press.

—— (1993) 'Palestinians in Israel: changes in the juristical status and political power' *Arab Studies: A Monthly, Cultural, Economic and Social Review*, May–July, 17–45.

Jirys, S. (1976) *The Arabs in Israel*, New York: Monthly Review.

Kamin, C. (1987) 'After the catastrophic: The Arabs in Israel – 1948–51', *Middle Eastern Journal* 23: 453–495.

Kazi, H. (1987) 'Palestinian women and the national liberation movement: a social perspective', in Khamsin Collective (ed.) *Women in The Middle East*, London and New Jersey: Zed Books, 26–39.

Kreiter, H. and Kreiter, S. (1978) *Cognitive Orientation and Behavior*, New York: Springer.

Lustick, I. (1980) *Arabs in the Jewish State: Israel's Control of a National Minority*, Austin, TX: Texas University Press.

Mar'i, M. (1983) 'Sex role perceptions of Palestinian males and females in Israel', Unpublished Ph.D Dissertation, East Lansing, MI: Michigan State University.

—— and Mar'i, S. (1985) 'The role of women as change agents in Arab society in Israel', in M. Safir, M. Mednick, D. Izraeli and J. Bernard (eds) *Women's World: From the New Scholarship*, New York: Praeger, 251–259 (in Hebrew).

Nath, K. (1978) 'Education and employment among Kuwaiti women', in L. Beck and N. Keddie (eds) *Women in the Muslim World*, Cambridge, MA: Harvard University Press, 172–188.

Nevo, J. (1989) 'Attack on opressed women: examination of testimonies from the field', Unpublished paper presented in the national criminological conference in May 29 – Jerusalem: Hebrew University.

Nimer, O. (1971) 'The Arab women in Israel', *Values* 3, 15: 44–47 (in Hebrew).
Peled, Y. (1992) 'Ethnic democracy and the legal construction of citzenship: Arab citizens of the Jewish state', *American Political Science Review*, 86, 2: 432–443.
Prothro, E. and Diab, L. (1974) *Changing Family Patterns in Arab East Beirut*, Beirut: American University Press.
Rishmawi, M. (1988) 'The legal status of Palestinian women in the occupied territories', in N. Toubia (ed.) *Women of the Arab World*, London and New Jersey: Zed Books, 79–92
Senker, C. (1989) *Palestinian Women in the Uprising: The Israeli Mirror*, London.
Shaloufeh Khazan, F. (1991) 'Change and mate selection among Palestinian women in Israel', in B. Swirski and M. Safir (eds) *Calling the Equality Bluff: Women in Israel*, New York: Pergamon Press, 82–89.
Smock, A. and Yousef, N. (1977) 'Egypt: from seclusion to limited participation', in J. Giele and A. Sinck (eds) *Women's Role and Status in Eight Countries*, New York: John Wiley & Sons,
Touma, E. (1981) 'Liberation of Arab women not sexual crisis', *Haifa Al-Jadeed*, 12, December (in Arabic).
Warnock, K. (1990) *Land before Honour: Palestinian Women in the Occupied Territories*, New York: Monthly Review Press.
Yisraeli, E. (1980) 'Adult education in the Arab Druze sector', *Studies in Education*, 25: 139–154 (in Hebrew).
Zureik, E. (1988) 'The Palestinians in Israel: a study in internal colonialism', R. Khalidi (ed.) *The Arab Economy in Israel*, London: Croom-Helm, 112–146.

7

HOMEFRONT AS BATTLEFIELD
Gender, military occupation and violence against women

Simona Sharoni

> There is a strong connection between violence against women and violence in the Occupied Territories. A soldier who serves in the West Bank and Gaza Strip and learns that it is permissible to use violence against other people is likely to bring that violence back with him upon his return to his community.
>
> (Ostrowitz in interview with Sharoni 1990)

INTRODUCTION

In April 1989 Gilad Shemen, a twenty-three-year-old Israeli-Jewish man doing his military service in Gaza, shot and killed a seventeen year-old Palestinian woman, Amal Muhammad Hasin, as she was reading a book on her front porch. The Regional Military Court convicted Shemen of carelessness in causing Hassin's death, but he was released after an appeal. Two years later, on June 30, 1991, Gilad Shemen shot and killed his former girlfriend, nineteen-year-old Einav Rogel.

In an interview right after her death, Einav Rogel's parents recalled that their daughter had supported Gilad Shemen unconditionally during his military trial, trying to convince everyone around her that he was not guilty. Yet during that entire period Einav did not tell anyone that Gilad also had been violently abusing her. She did not recognize the connections between Gilad's shooting of a Palestinian woman and the violence and fear that Gilad brought to her own relationship with him. Einav Rogel lived and died in a society that draws clear distinctions between 'us' and 'them,' and usually doesn't even record the names of Palestinians who are shot. At the same time, she did not realize that, like many other Israeli women and most Palestinians (both women and men) in the West Bank and Gaza Strip, she belonged to a high risk population – since she lived in the line of fire of an Israeli man who had learned to use his gun to deal with crises and difficult situations.

This tragic story underscores the complex relationship between sexism, militarism, and violence against women. This relationship has been explored at length by feminist scholars and activists (Accad 1990; Morgan 1989; Jeffords 1989; Ruddick 1989; Cooke 1988; Enloe 1988; Cohn 1989; Reardon 1985). It is extremely important, however, also to situate the tragedy of Gilead Shemen within the specific sociopolitical context of the third decade of Israeli occupation of the West Bank and Gaza Strip.

This chapter deals with the impact of the Israeli Occupation of the West Bank and Gaza Strip on women's lives, by highlighting the connections between, on the one hand, the social construction of gender identities and gender relations in Israel and, on the other, the use of violence in the Occupied Territories and on the Israeli 'homefront.' Although the chapter focuses primarily on the origins and manifestations of men's violence against women, it does not treat violence as a set of practices with which men are born; these practices are used rather as a means of coping and they are acquired by and reinforced in Israeli men through education and social interaction. The chapter will critically examine the dominating role of the Israeli military in all spheres of Israeli society, and the social and political implications of militarization and violent conflict for women's lives both in Israel and in the West Bank and Gaza Strip. The central argument here is that violence against women is intimately related to other forms and practices of violence, especially in the context of the Israeli Occupation of the West Bank and Gaza Strip.

Much of the recent literature on women's ways of coping with the violence inflicted upon their lives by war and conquest tends to remain caught between two opposed, stereotypical images: the image of woman-as-victim; and that of woman-as-heroine (Sharoni 1993; Abdo 1991). To move beyond the constraints of these dominant representations, this chapter will focus on particular stories which demonstrate the range of women's daily experiences of violence and the diverse strategies which women have employed in resisting violence. In addition to drawing attention to the multifaceted struggles of Palestinian women in the Israeli Occupied West Bank and Gaza Strip, the chapter will explore recent attempts by Israeli-Jewish feminists and peace activists to connect, on the one hand, their own resistance to the violence inflicted upon Palestinians in the West Bank and Gaza Strip by the Israeli military and, on the other hand, their own struggles to end male-inflicted violence in the lives of women in Israel.

MILITARY OCCUPATION: IMPLICATIONS FOR WOMEN'S LIVES

The pitchforks in local women's hands are brown:
nails and nails,
rust and rust in their edges
and a long wooden handle

> intended to pierce
> the flesh of our faces,
> to pluck.
> Our women,
> pluck their eyebrows.
> (Sternfeld 1988, quoted in Ben Ari, 1989: 375)[1]

This poem dramatizes the intersection between sexist, militaristic, and racist discourses. Three particular distinctions are at play here: between men and women; between 'us,' the local-patriots, and 'them' (the 'enemy'); and between 'our' women and 'their' women. These configurations reflect particular power relations which are grounded in and reinforced by the reality of military occupation. The poem calls attention to the fact that Israeli soldiers have used the pretext of cultural and moral superiority to justify their use of excessive power over Palestinians, both women and men, in the West Bank and the Gaza Strip. While he treats both Palestinian and Israeli women as objects with no political agency, Sternfeld maintains a clear distinction between them by using Orientalist depictions of Palestinian women.[2] Moreover, Sternfeld uses his portrayal of Palestinian women as vicious enemies, ready to pierce and pluck the flesh of Israeli soldiers' faces, to resolve the tension between the Israeli army's self-portrayal as a humane army that has tried at all costs to prevent women and children from suffering (Gal 1986) and the reality of military occupation which has been sustained through an indiscriminate use of violence against Palestinians as a whole, including women and children.

The indiscriminate use of violence and oppressive practices against Palestinians in the West Bank and Gaza Strip, especially since the outbreak of the *intifada* in 1987, has had particular implications for women's lives. Palestinian women have had to confront violence on two intimately related fronts: as members of the Palestinian community; and as women. Their homes and bodies have become the battlefields for these confrontations.

Rita Giacaman's and Penny Johnson's gendered examination of the first year of the *intifada* (Giacaman and Johnson 1989) highlights these multi-dimensional confrontations in relation to the ongoing struggles of Palestinians against the Israeli Occupation. Giacaman's and Johnson's retelling of the story of Umm Ruquyya ('mother of Ruquyya'), the mother of a young Palestinian woman activist in the West Bank, captures the harsh reality of life for women under occupation: 'we went to visit her on November 6, 1988 when we heard that the family house had been demolished by the Israeli army. Hers was one of more than 100 houses destroyed in the northern Jordan Valley, leaving about 1,000 persons homeless and devastated' (Giacaman and Johnson 1989: 155). Ruquyya was the first among both the men and the women of her community to mobilize resistance in response to the demolitions of family homes by the Israeli military, thus serving as 'one of many women forging a new chapter in the history of the Palestinian women's movement' (Giacaman and Johnson 1989).

The Israeli military has used multiple strategies to suppress the unprecedented political mobilization of Palestinian women. During the first two years of the *intifada*, the Israeli military used tear gas, which was found to cause miscarriages, to suppress demonstrations and to deter women from future participation in public political events. In addition, by declaring the Palestinian Women's Working Committees and any other form of social and political organizing by women illegal, the Israeli military authorities created a pretext for massive arrests of Palestinian women. Women were arrested and interrogated not only because of their political activities but also in order to put pressure on their families and to get incriminating evidence against family members (Strum 1992). Sexual harassment and sexual violence, in addition to other means of torture and humiliation, have also been used as weapons against Palestinian women (Rosenwasser 1992; Strum 1992).

To live under military occupation is to live in a permanent state of war, with no place to hide and no cease-fires. Palestinians have lived with the oppressive and violent reality of occupation since 1967. Only with the outbreak of the *intifada* in December 1987 have these circumstances begun to be exposed and subjected to public scrutiny, as abuses such as sexual harassment and sexual violence against Palestinian women have been added to the agenda of human rights and women's peace groups in Israel. Since the beginning of the *intifada*, the Women's Organizations for Women Political Prisoners (WOFPP) in Tel Aviv and Jerusalem have received numerous complaints of sexual violence committed by Israeli military forces against Palestinian women in the Occupied Territories. Such incidents occur not only during interrogation but also in connection with street patrols and the suppression of demonstrations.

The case of thirty-six-year-old Fatma Abu Bacra from Gaza, who was arrested in November 1986, is a representative example of the sexual abuse and humiliation which Palestinian women have suffered during interrogation by the Israeli Security Services. One Israeli male interrogator touched her face and breast, while another showed her a picture of a naked man and told her that the picture was of himself. He then took off his clothes and threatened to rape her. Abu Bacra reported the torture to her male lawyers, but only submitted a detailed affidavit about the sexual abuse later on when she had a woman lawyer. In this affidavit, Fatma Abu Bacra describes how, further, she was removed by one of her interrogators to a separate room, with no policewoman present (in violation of regulations), and forced to sit in a corner with her head wedged between the interrogator's legs while he touched her, verbally abused her, threatened her with rape, and eventually reached sexual climax (WOFPP Report 1992).

This affidavit was accepted on November 22, 1988 by a military judge, as the basis for a pre-trial hearing on the validity of admissions which Abu Bacra had made under sexual torture. In the spring of 1989 the pre-trial hearings began, and during the proceedings a plea bargain was reached: under pressure Abu Bacra agreed not to challenge the way in which her confession had been obtained; in return, she was promised that her sentence would not exceed five years. However, in June 1989 Fatma Abu Bacra was given a seven-year sentence. She appealed and the sentence was reduced to six years. The Israeli authorities later used this unfulfilled bargain to claim that Abu Bacra had retracted her statement on the torture and sexual abuse which she had undergone during interrogation. Since the minutes of these proceedings remained classified, Abu Bacra's lawyer appealed to the High Court of Justice demanding the right to publish them. Finally, in order to circumvent publication of the interrogation minutes the authorities decided to release Fatma Abu Bacra a year earlier than her sentence had stipulated, in November 1991, on the condition that her appeal be withdrawn (WOFPP Report 1992).

Although the Palestinian uprising has not broken the silence and denial of Israeli society in general, it has served as a turning point in the political awareness of many Jewish women in Israel. The Israeli Women's Organizations for Women Political Prisoners, for example, have been documenting particular cases, like Fatma Abu Bacra's, of torture and sexual violence experienced by Palestinian women; thus they have, like other Israeli women's peace groups, begun to expose the connections between the excessive use of sexual violence as a weapon against Palestinian women, the sharp increase in violence against women in Israel, and the politics of the Israeli–Palestinian conflict.

Especially since the outbreak of the *intifada*, many women in Israel have begun to challenge the marginal, passive roles assigned to them in Israeli society and politics. For the first time in the history of the state, women have organized and taken clear positions against state and military policy. Israeli women have voiced strong dissent against the Occupation and against the brutal violence used by Israeli soldiers against Palestinian civilians in the West Bank and Gaza Strip. New women's protest groups such as Women in Black, *Reshet* (The Israeli Women's Peace Net), Women's Organizations for Women Political Prisoners, *Shani* – Women Against the Occupation, and The Women and Peace Coalition have emerged, providing opportunities for women to step out of their socially assigned, politically peripheral roles (Sharoni 1993a; Deutsch 1992; Chazan 1991).

Israeli women's political interventions have not found widespread acceptance among Israeli men. Women in Black groups throughout Israel have become targets for verbal and physical violence that is almost always laced with sexual innuendo. The epithets which some men shout at women protesters – 'whores of Arafat' or 'Arab lovers' – reflect the culture of militarism and sexism within which Israeli men are socialized. But while many Israeli men find it difficult to understand what has motivated women to protest weekly for more than four years against the Occupation, for Women in Black the interconnectedness between militarism and sexism remains a tangible, experienced part of their struggle to find a political voice for themselves and to express their opposition to the continuing occupation of the West Bank and Gaza Strip.

The image of the brutal occupier who commits daily violence against Palestinian women and children and brings the violence home to his family and friends does not fit the national image of the brave Israeli soldier who has no choice but to fight in order to protect Israeli women and children. But, slowly, the message is starting to become clear: the violent patterns of behavior that are used by the Israeli army against Palestinians in the West Bank and Gaza Strip are part of a culture of unchallenged sexism, violence, and oppression which women face daily on Israeli streets and in their homes.

EVERY WOMAN IS AN OCCUPIED TERRITORY

> They bombarded us
> they shoot one salvo after another
> Directing toward us
> strafing and guns.
> Nurit,
> I have encircled the Third Battalion,
> Now I want
> to encircle you.[3]

In this 'love' poem, the heroic/erotic discourse fuses militaristic metaphors as expressions of love and lust, violence and sex. Women – like the enemy – are to be encircled and occupied by Israeli heroes.[4] This particular poem is but one representation of the perverse relationship between militarism and sexism that surfaces in most spheres of Israeli society. That relationship is clearly inscribed in the Hebrew language as well. The multiple meanings of the word *kibush* represent a striking case in point. The word *kibush* is the most commonly used Hebrew term for the Israeli occupation of the West Bank and Gaza Strip – and is also used in Hebrew to describe conquest either of a military target or of a woman's heart. This conflation of women and military targets is not merely linguistic, but rather informs numerous practices in Israeli society in general and in the Israeli military in particular. During military training exercises, for example, the strategic targets are quite often named after significant women in the soldiers' lives: women, like military targets, must be protected so that they will not be conquered by the 'enemy'; while men must fight, occupy, and protect. These examples suggest interplay between gender, language, and politics in Israel, which has been grounded in and reinforced by particular social and political conditions.

Israeli men soldiers have constantly to prove their readiness to sacrifice their lives on the battlefield; while Israeli women are left with no other choice but to 'sacrifice' their lives, freedom, and independence on the homefront. Israeli women's bodies, hearts, and identities have been conquered, 'occupied', and objectified in numerous ways. Language further reflects the state of Israeli gender relations and cultural politics. The common word for 'husband' in Hebrew, for example, is *baal*, which also means both 'owner' (noun) and 'had intercourse with' (verb), indicating that women are perceived as their husbands' property. Israeli men's 'private' ownership of 'their' women has in fact been extended to the state.

The treatment of Israeli women as occupied territories also manifests itself in numerous practices of control over women's identities, roles, and bodies, which have been reinforced by the escalation of the Arab–Israeli/Israeli–Palestinian conflict. Three practices are of particular relevance here: the steep rise in violence against women in Israel since the outbreak of the *intifada*, and particularly in the aftermath of the Gulf War; the mobilization of women's reproductive work in the service of the state under the pretext of 'demographic war', a pretext that has been used to justify impediments on women's reproductive rights in general and restrictions on abortion laws in particular; and the mobilization of gender identities in service of the state.

Violence against women

The connections between sexism and militarism, and between violence against the 'enemy' on the battlefield and against women on the homefront, are by now considered old feminist themes (Enloe 1988; Woolf 1977; Brownmiller 1975). Women peace activists in Israel have become particularly aware of these connections since the outbreak of the *intifada*. Rachel Ostrowitz, editor of the Israeli feminist magazine *Noga*, calls attention to the similarities between the ways in which both Palestinians and women are treated by Israeli men:

> The similarity in the treatment of oppressed human beings is clear to us. When we read every day about nameless dead Palestinians, we remember that women are often treated as persons without names. 'Women are all the same,' they tell us; 'all Palestinians are the same.' The voices merge.
>
> (Ostrowitz 1989: 14)

The dehumanization of both Palestinians and women legitimizes the discrimination, the humiliation, and the oppression and violence inflicted daily upon them. Rachel Ostrowitz further delineates the connections between the use of violence against Palestinians in the West Bank and Gaza Strip and the steep increase in violence against Israeli women on the homefront:

> Oppression is oppression is oppression . . . There is a strong connection between violence against women and violence in the Occupied Territories. A soldier who serves in the West Bank and Gaza Strip and learns that it is permissible to use violence against other people is likely to bring violence back with him upon his return to his community. This has direct implications for our lives as women.
> (Ostrowitz in interview with Sharoni 1990)

The structures which Ostrowitz sees as responsible for the oppression and humiliation suffered by Palestinians and by Jewish women are grounded in and reinforced by the unchallenged acceptance of 'national security' as the top priority in Israel:

> The twisted priority that land is more important than human life reminds us of other twisted priorities – military equipment instead of equal pay for women, or better education for the future generation .
> (Ostrowitz 1989: 14–15)

Such 'twisted priorities' have served as pretexts in the recruitment of women's bodies in the service of the state.

Women's bodies as national battlefields

In virtually all societies, the military maintains a major role in the shaping of gender identities and gender relations, especially in war zones. Focusing on the army as a major 'agent of socialization' for men, Cynthia Enloe points out how the juxtapositions of masculinity against femininity and of men against women serve as important ideological frameworks in the military:

> Military forces past and present have not been able to get, keep and reproduce the sorts of soldiers they imagine they need without drawing on ideological beliefs concerning the different and stratified roles of women and men. Without assurances that women will play their 'proper' roles, the military cannot provide men with the incentives to enlist, obey orders, give orders, fight, kill, re-enlist, and convince their sons to enlist. Ignore gender – and social construction of 'femininity' and 'masculinity' and the relations between them – and it becomes impossible adequately to explain how military forces have managed to capture and control so much of society's imagination and resources.
> (Enloe 1988: 212)

Enloe's powerful critique demonstrates how women, and strategies for controlling women, have been used to support military campaigns around the world. This has been definitely true in the case of Israel.

In direct relation to men's wars on the battlefield, Israeli Jewish women have been 'recruited' on more than one front. Since the early 1950s, Israel has utilized one myth in particular – that of a nation under siege – to justify political practices such as the 'demographic war.' Prime Minister David Ben Gurion actually raised the issue of women's fertility to the level of national duty, arguing that: 'increasing the Jewish birthrate is a vital need for the existence of Israel, and a Jewish woman who does not bring at least four children into the world is defrauding the Jewish mission' (quoted in Hazleton 1977: 63). In the 1980s, that old myth was once again invoked to fit the political agendas of the time. The 'Efrat Committee for the Encouragement of Higher Birth Rates' linked the public debate on abortion to the 'demographic war,' for which women's bodies had served as the designated turf. Utilizing the rhetoric of religious anti-abortion groups, the Efrat Committee called upon Jewish women to fulfill their national duty by bearing more children in order to replace the Jewish children killed by the Nazis (Yuval-Davis 1987).

An extreme example of how this ideology was put into practice was a narrowly defeated proposal by the then advisor to the Minister of Health, Haim Sadan. Sadan proposed to force every Jewish woman considering abortion to watch a slide show which would include, in addition to horrors such as dead fetuses in rubbish bins, pictures of dead Jewish children in the Nazi concentration camps (Yuval-Davis 1987). This shocking example is not unique. In fact, the Holocaust has been mobilized in this way by the state and its dominant institutions not only to justify hardline political positions and military campaigns, or racist and sexist policies such as 'demographic war', but also to clearly mark the borders between what it means to be a woman and what it means to be a man in the Jewish state.

Gender identities in the service of the state

In order to have a place in the Israeli collectivity and to share the patriotic ethos of 'national security,' Israeli women have to enter the narrow doorways marked 'mother' or 'wife' through their affiliation with a male soldier. Former Knesset Member Geula Cohen, the founder of the extreme right-wing Tehiya party, utilized the rhetoric of national security and her platform as a political woman to remind Israeli women of this national obligation: 'the Israeli woman is a wife and a mother in Israel, and therefore it is her nature to be a soldier, a wife of a soldier, a sister of a soldier, a grandmother of a soldier. This is her reserve duty. She is continually in military service' (Hazleton 1977: 63). It is in this light, according to former Israeli Knesset Member Marsha Freedman, that women's liberation in Israel is deemed a threat to 'national security' (Freedman 1990: 108).

In addition to the 'reserve duties' articulated by Geula Cohen, Israeli women have to mediate the relationship between Israeli male soldiers and their motherland. The word for 'homeland' in Hebrew is *moledet*, which is a feminine noun derived from the verb 'to give birth.' Moreover, 'homeland' is almost always presented in Israeli popular culture as motherland, and men are portrayed as sons who return home to the warmth, love, and support of their beloved mothers. But Israeli men are socialized to understand that in order to be worthy of homecoming they must accept the need to sacrifice their lives for the homeland as a national duty and an honor. The national narrative of heroic sacrifice is constituted from early childhood onward through mythologized stories such as these of *Masada* and *Tel-Hai*, and becomes the major model for measuring loyalty to the state and its ideology (Zerubavel 1990, 1991). This erotic/patriotic complex informs politics not only on the battlefield but on the homefront as well. For example, the funerals of Israeli soldiers are usually broadcast on radio and TV, and become politically charged as top government officials are shown comforting weeping mothers and commending them for raising up sons who were ready to sacrifice their lives for their homeland.

In sum, the institutionalization of Israeli Jewish women's roles as the primary caretakers of a nation of soldiers would not have been possible without certain dominant interpretations of the Israeli–Palestinian and the Arab–Israeli conflict. Similarly, the 'recruitment' of Israeli women's reproductive organs into the service of the state depends upon the prevailing myth of Israel as a nation under siege, the underpinning of political practices such as the 'demographic war.' The linkages drawn between the Holocaust and Israel's anti-abortion campaigns further reinforce a particular order of gender relations in Israel, revolving around militarized men who fulfill the 'sacred' task of protecting women and children on the 'homefront.'

ZIONISM, 'NATIONAL SECURITY' AND THE CONSTRUCTION OF MILITARIZED MASCULINITY

The centrality of the military among Israeli social and political institutions has often been taken for granted. However, this centrality is not natural. It has been constituted and reinforced through specific ideologies and practices. The establishment of the Israeli state and the elevation of its hegemonic Zionist ideology made 'national security' a top priority, designed to secure the survival not only of the country but of the Jewish people at large. This view of the priority of Israel's 'national security' is grounded in a particular historical narrative concerning the birth of Israel, an histroical narrative whose core is formed by several unchallenged myths. The notion of Palestine as the 'land without people for the people without land', and the claim that the Zionists welcomed the partition of Palestine while the Palestinians rejected it belong to one sort of myth. Another is that the

Palestinians fled Palestine in 1948 despite Jewish leaders' efforts to get them to stay, or that after the 1948 war Israel extended its hand in peace to all neighboring countries, but not a single Arab leader responded (Segev 1986; Flapan 1987; Morris 1988).

Through such self-legitimating myths, the state's dominant historical narrative regarding the birth of Israel has hardened into an ideological shield that has been projected on to Israeli society as well as the Jewish Diaspora. The Israeli military has become the major agent for facilitating this process of ideological projection. Since the establishment of the state, the declared objectives of the Israeli doctrine of 'national security' have always been to build a cohesive, unified front. Accordingly, as discussed earlier, Israel's dominant conceptualizations of 'national security' have been constructed around unchallenged representations of Israel as 'a nation under siege,' surrounded by enemies that threaten to throw the entire population into the sea; and this myth has been reinforced through constant invocations of the Holocaust and through political manipulations of facts concerning the 1948, 1967 and 1973 wars. In recent years, a new generation of Israeli historians such as Benjamin Beit-Hallahmi (1992), Simha Flapan (1987), Benny Morris (1988), Anita Shapira (1992), and Tom Segev (1986) have begun to challenge the conventional belief that in all the wars it has fought Israel's action have been just and inevitable, guided by the principles of human dignity, justice, and equality. However, despite compelling evidence presented in this scholarship, most Jews in Israel and in the Diaspora still cling to the illusion that Israeli domination and repression have been inevitable; essential to the survival and security of the nation. Simha Flapan reflects on the rigidity of these myths and their centrality in Israeli society and politics by sharing his own experience:

> Like most Israelis, I had always been under the influence of certain myths that had become accepted as historical truth. And since myths are central to the creation of structures of thinking and propaganda, these myths had been of paramount importance in shaping Israeli policy.
>
> (Flapan 1987: 8)

Yet, what remains missing even from revisionist accounts, like Flapan's, of Israeli history is a gendered understanding of dominant historical narratives and myths, exploring the particular conceptions of masculinity and femininity that these historical narratives present.

By making 'national security' a top priority, by grounding it in specific interpretations of Zionist ideology and of the history of the Jewish people, and by turning military service into a national duty, the state has offered Israeli Jewish men – especially those of European or North American descent – privileged status in Israeli society. Furthermore, since one of the primary objectives of the Israeli doctrine of 'national security' has always been to build a cohesive, unified front, 'national security' has been used to justify Israeli militaristic and expansionist policies and political practices – and also to neutralize and thus to legitimize and reinforce existing inequalities among Israel's citizens along lines of gender, ethnicity, class, and political affiliation. Israel's disenfranchised populations have in effect been asked to understand that until the Arab–Israeli conflict is resolved they must stand united against the external 'enemy' (Swirski 1989; Shohat 1988). Attempts by grassroots social movements representing Israel's second-, third-, and fourth-class citizens – women; Jews from Arab and North African countries; and Palestinians who hold Israeli citizenship – to protest against discriminatory state policies have been dismissed under the premise of 'national security.' What remains particularly concealed in most existing critiques of Israeli 'national security' is the fact that the rhetoric of 'national security' depends on the preservation of the status quo not only with respect to Israeli–Palestinian/Arab–Israeli conflicts but also with respect to the social construction of gender identities and roles. Israeli feminists and activists are gradually coming to terms with the ways in which the construction of Israeli masculinity is linked to the militarized political climate in Israel and in the region. More specifically, many women peace activists in Israel have recently argued that the institutionalization of 'national security' as a top priority in Israel contributes to gender inequities and legitimizes violence against Palestinians and against women.

The Israeli state's doctrine of 'national security' depends both upon men who are ready to serve as soldiers, as fighters on the battlefield, and upon women who are ready to adjust to the needs of the Israeli collective experience. On one hand, women are socialized into the roles of unconditional supporters, exceptional caretakers, and keepers of the homefront; on the other hand, they are expected to remain vulnerable and in need of protection. While these contradictory messages no doubt result in major problems in the construction of Israeli women's identities, Israeli popular culture has attempted to resolve the contradiction by subordinating both roles to the primacy of national identity and by utilizing both images as pretexts for wars.

It is important to note that the practical and symbolic mobilization of gender identities, roles, and bodies in the service of the Jewish state would not have been possible without engaging the mythologies of Israel as 'a land with no people for the people with no land,' as the only safe place for Jews in the aftermath of the Holocaust, and as 'a nation under siege.' These narratives, sanctioned by some of the major tenets of Zionist ideology, have been used to justify the masculine and militaristic practices associated with the establishment of the state of Israel; through them Israel's reassertion of 'masculinity' has been explained in terms of the need to end a history of weakness and suffering.

The symbol of the *sabra* can stand as an exemplary metaphor for this reassertion of masculinity. Named after the indigenous cactus fruit, *sabar*, which is tough and prickly on the outside and soft and sweet on the inside, the image of the *sabra* has played an important role in the construction of the identity of the new generation of Jews born in Israel. This generation has been portrayed as the antithesis of the weak, persecuted, and helpless Jews most commonly associated with collective traumatic memories of the Holocaust. The image of the *sabra* as the antithesis of the Diaspora Jew is used to reinforce the notion that Israel's offensive operations and military campaigns are a matter of national survival (Sharoni 1992a); in turn, the *sabra*'s offensive and aggressive codes of behavior are justified through the ahistorical appropriation of the motto of 'never again.' Yet, exploitation of the *sabra* image is grounded not only in the juxtaposition of the image of the *sabra* against the image of the persecuted Jew in the Diaspora, but also in the juxtaposition of masculine and feminine identities. In the terms of this gendered juxtaposition, men must be offensive on the battlefield in order to protect vulnerable women on the homefront. The underlying model of relations between strong, possessive men and weak, helpless women serves not only as a pretext for continued male domination on the homefront and as justification for the use of violence on the battlefield but also, more generally, as justification of violent behavior by men. Thus the dominant juxtapositions of the invincible *sabra* man with the weak and helpless Diaspora Jew, and of men as protectors and women as needing protection, have been strongly informed – and reinforced; even justified – by Zionist ideology and by the unchallenged centrality of 'national security' in Israel.

The *sabra* has become a common metaphor in Israeli literature and popular culture for Israeli men, who are thus characterized as strong and brave, pragmatic, aggressive, and emotionally tough. Few have noticed, however, that only the tough and prickly outside part of the cactus fruit has been incorporated into readings of this metaphor. There are no references in Israeli popular culture to the soft and sweet inside part of the fruit, which might be deemed 'feminine.' The *sabra* metaphor may therefore shed light on the ambiguities embedded in Israeli society's expectations of women. On one hand, when Israeli men are on the homefront, women are relegated to conventionally gendered roles: they have to be 'inside,' 'soft,' 'tender' and 'sweet.' On the other hand, during wartime when men are on the battlefield, women are expected to step out of their traditional roles and to enter, if only temporarily, the public political arena. During such periods pragmatic, assertive, and tough behavior on the part of women is praised as a significant contribution to the collective national effort.

CONCLUDING REMARKS

This chapter explored the social construction of gender identities and gender relations in Israel in the context of the Israeli–Palestinian conflict and, especially, in relation to the third decade of Israeli Occupation of the West Bank and Gaza Strip. The chapter focused primarily on the relationship between sexism and militarism and its implications for Israeli and Palestinian women's lives, and on the connections between violence on the 'battlefield' and violence on the 'homefront.'

The murders of Amal Muhammad Hasin and Einav Rogel by Gilad Shemen are but one symptom of the strong link between militarism and sexism; sexual abuse and violence used against Palestinian women political prisoners is another manifestation of this pathological relationship. However, Israeli society on the whole has so far refused to address the interconnectedness of militarism and sexism; in particular, it has ignored the relationship between the escalation of violent practices by Israeli soldiers against Palestinians in the Occupied Territories and the steep increase in men's violence against women in Israel.

Journalist Gabi Nizan was among the few Israelis who have tried to situate the murders of Einav Rogel and Amal Muhammad Hasin in the social and political context of military occupation. A few days after Einav Rogel's murder he wrote in the Israeli mass circulation newspaper *Hadashot*:

> In a country without wars, Einav Rogel and Amal Muhammad Hasin could have been good friends. In such a world Gilad Shemen could have been a good friend of both of them. But in our society, Shemen met both of them with a gun in his hand. This is very normal for an Israeli his age and it is normal that a gun shoots. This is what weapons are for.
> (*Hadashot*, July 4, 1991: 16)

Gilad Shemen will probably be sent to a mental health institution and not to jail; other Israeli men like him will continue to use violence as a means of dealing with problems both on the battlefield and on the 'homefront.' At the same time even the more liberal sectors of Israeli society hesitate to link publicly the use of violence against Palestinians in the West Bank and Gaza Strip with the increase in violence against Israeli women at home. When the Israeli media finally took note of the tremendous increase in incidents of violence against women – including murder – over the years of the Occupation, the reports lacked any reference to the broader historical and political context within which such incidents of violence emerge and are tolerated. There has been hardly any mention of the impact of the Gulf War on the masculine self-image and national identity of Israeli men, or on the increasing vulnerability of Israeli women.

The Gulf War was the first time that Israeli men were not drafted during wartime. Men remained on the 'homefront,' confronted with their families' fears, with their own fears, and with the vulnerability and helplessness of being locked in a sealed room. The image of the invincible Israeli soldier ready at all costs to protect women and children was endangered. Israeli men became increasingly uncomfortable with this unfamiliar role; many used the word 'impotent' to describe their feelings. Unable to express themselves violently against Arabs, as they have been trained and conditioned to do, and confronted with the fact that the separation between violence on the 'battlefield' and violence on the 'homefront' existed only in their minds, many Israeli men 'cured' their feelings of 'impotence' and longings for the excitement of the battlefield by projecting their aggression on to women (Sharoni 1991).

Separating one set of inequalities from another reduces possible threats to the often unchallenged regimes of power and privilege. But such connections do, none the less exist. Many of the same Israeli men who carry out violent practices against Palestinian men and women in the Occupied Territories with an official license from the state treat the significant women in their life as their 'occupied territories.' The murders of Amal Muhammad Hasin and Einav Rogel by the same man in military uniform is not a tragic coincidence, but a direct result of the Israeli–Palestinian conflict. In a context where every man is a soldier, every woman becomes an occupied territory.[5]

Feminist scholars and activists who are committed to social change need to challenge the silences and gaps in the conventional scholarship on the Israeli–Palestinian conflict, and to examine further the relationship between militarism, violence, and the social construction of gender in Israel and elsewhere.

NOTES

1 This poem appeared first, in Hebrew, in 'Intifada Diary,' *Iton 77*, 106–7, 93–95, 1988 and was later cited in Ben Ari, 'Masks and Soldiering' 1989: 375. Translation into English by E. Ben Ari.
2 'Orientalism' refers to a view of the 'Orient' as antithetical to and radically different from the 'West.' For an in-depth examination of the social and political implications of Orientalism see E. Said, *Orientalism*, New York, Vintage Books, 1979. For an excellent discussion of these questions as they are represented in Israeli cultural politics see E. Shohat, *Israeli Cinema: East/West and the Politics of Representation*, Austin, University of Texas Press, 1989. For more information on the representation of Middle Eastern women see Z. Hajaibashi, 'Feminism or Ventriloquism: Western Presentation of Middle East Women,' *Middle East Report*, 1991, issue 172, pp. 43–45, and Judy Marbo, *Veiled Half Truths: Western Travellers' Perceptions of Middle Eastern Women*, New York, St Martin's Press, 1991.
3 This poem is included in the preface of Meir Shapira's 1973 war journal *Written in Battle: 'A Battle's Journal of the Yom Kippur War*,' Tel Aviv: Alef Publishing House 1976. The poem appears after the author's short introduction along with two other war poems written by Chaim Nachman Bialik and Natan Alterman, icons of Israeli poetry. The term 'the Third Battalion' refers to an Egyptian military battalion that participated in the 1973 war. This war journal was published in Hebrew; the English translation is mine.
4 The reference in the poem to women as territories to be occupied and as military targets echoes the interconnectedness between sexism and militarism in Israel. For more on this topic see my article: 'Every Woman is an Occupied Territory: The

Politics of Militarism and Sexism and the Israeli–Palestinian conflict,' *Journal of Gender Studies*, Vol. 1, 3: 447–462, 1992 (Special Issue: Gender and Nationalism).
5 This reference to women as occupied territories is discussed in R. Metzger, *The Woman Who Slept with Men to Take the War Out of Them*, Culver City, CA, Peace Press, 1985.

BIBLIOGRAPHY

Abdo, N. (1991) 'Women of the *Intifada*: gender, class, and national liberation', *Race & Class* 32, 4: 19–34.

Accad, E. (1990) *Sexuality and War: Literary Masks of the Middle East*, New York: New York University Press.

Beit Hallahmi, B. (1992) *Original Sins: Reflections on the History of Zionism and Israel*, Concord, MA: Pluto Press.

Ben-Ari, E. (1989) 'Masks and soldiering: the Israeli army and the Palestinian uprising', *Cultural Anthropology*, 44: 372–389.

Brownmiller, S. (1975) *Against Our Will: Men, Women and Rape*, New York: Simon and Schuster.

Chazan, N. (1991) 'Israeli women and peace activism', in B. Swirski and M. Safir (eds) *Calling the Equality Bluff: Women in Israel*, New York: Pergamon Press.

Cohn, C. (1987) 'Sex and death in the rational world of defense intellectuals', in L. Forcey (ed.) *Peace: Meanings, Politics, Strategies*, New York: Praeger. Originally appeared in *Signs: Journal of Women in Culture and Society*, 12, 4, (1987): 687–718.

Cooke, M. (1988) *War's Other Voices: Women Writers on the Lebanese Civil War, 1975–1982*, Cambridge and New York: Cambridge University Press.

Deutsch, Y. (1992) 'Israeli women: from protest to a culture of peace', in D. Hurwitz (ed.) *Walking the Red Line: Israelis in Search of Justice for Palestine*, Philadelphia: New Society Publishers, 44–55.

Enloe, C. (1988) *Does Khaki Become You? The Militarization of Women's Lives*, London: Pluto Press.

Flapan, S. (1987) *The Birth of Israel: Myths and Realities*, New York: Pantheon Books.

Freedman, M. (1990) *Exile in the Promised Land: A Memoir*, New York: Fireband Books.

Gal, R. (1986) *A Portrait of the Israeli Soldier*, New York: Greenwood Press.

Giacaman, R. and Johnson, P. (1989) 'Palestinian women: building barricades and breaking barriers', in Z. Lockman and J. Benin (eds) *Intifada: The Palestinian Uprising against Israeli Occupation*, Boston: South End Press and MERIP, 155–169.

Hazleton, L. (1977) *Israeli Women: The Reality Behind the Myths*, New York: Simon and Schuster.

Jeffords, S. (1989) *The Remasculinization of America: Gender and the Vietnam War*, Bloomington and Indianapolis: Indiana University Press.

Morgan, R. (1989) *Demon Lover: On The Sexuality of Terrorism*, New York: W. W. Norton Co.

Morris, B. (1988) *The Birth of the Palestinian Refugee Problem*, Cambridge and New York: Cambridge University Press.

Ostrowitz, R. (1989) 'Dangerous women: the Israeli Women's Peace Movement', *New Outlook* 35, 6/7: 14–15.

Reardon, B. (1985) *Sexism and the War System*, New York: Teachers College, Columbia University.

Rossenwasser, P. (1992) *Voices from a Promised Land: Palestinian and Israeli Peace Activists Speak their Hearts*, Williamantic, CT: Curbstone Press.

Ruddick, S. (1989) *Maternal Thinking: Toward a Politics of Peace*, New York: Ballantine Books.

Segev, T. (1986) *1949: The First Israelis*, New York: The Free Press.

—— (1992) *The Seventh Million: The Israelis and the Holocaust*, Jerusalem: Domino Press (in Hebrew).

Shapira, A. (1992) *Land and Power*, Tel-Aviv: Am Oved (in Hebrew).

Sharoni, S. (1991) 'Silenced by war', in *New Directions for Women*, 20, 3: 1–4.

—— (1992) 'Women's alliances and Middle East politics: conflict resolution through feminist lenses', Paper presented at the Annual Meeting of the Association for Israel Studies (AIS), Milwaukee, WI, May.

—— (1992a) 'Militarized masculinity in context: cultural politics and social constructions of gender in Israel', Paper presented at the Annual Meeting of the Middle East Studies Association, Portland, OR, October.

—— (1993) 'Middle East politics through feminist lenses: toward theorizing international relations from women's struggles', *Alternatives*, 18, 1: 5–28.

—— (1993a) 'Conflict resolution through feminist lenses: theorizing the Israeli–Palestinian conflict from the perspective of women peace activists in Israel', Unpublished Ph.D dissertation, George Mason University.

—— (1994) *Gender and the Israeli–Palestinian Conflict: The Politics of Women's Resistance*, Syracuse, NY: Syracuse University Press (in press).

Shohat, E. (1988) 'Sepharadim in Israel: Zionism from the standpoint of its Jewish Victims', *Social Text*, 19/20, 1–35.

Strumm, P. (1992) *The Women as Marching: The Second Sex and the Palestinian Revolution*, New York: Lawrence Hill Books.

Swirski, S. (1989) *Israel: The Oriental Majority*, London and New Jersey: Zed Books.

WOFPP (1992) *Women's Organization for Political Prisoners Newsletter*, Tel Aviv.

Woolf, V. (1977) *Three Guineas*, London: Houghton Press, 1938; reprint, Middlesex: Pergamon Books.

Yuval-Davis, N. (1987) 'The Jewish collectivity', in Khamsin (ed.) *Women in the Middle East*, London and New Jersey: Zed Books.

Zerubavel, Y. (1990) 'New beginning, old past: the collective memory of pioneering in Israeli culture', in L. Silberstein (ed.) *New Perspectives on Israeli History: The Early Years of the State*, New York: New York University Press.

—— (1991) 'The politics of interpretation: Tel Hai in Israel's collective memory', in *The Journal of the Association for Jewish Studies*, 16, 1/2: 133–160.

8

TRENDS IN LABOR MARKET PARTICIPATION AND GENDER-LINKED OCCUPATIONAL DIFFERENTIATION[1]

Moshe Semyonov

The purpose of this chapter is to examine patterns in women's labor force participation and gender-linked occupational differentiation among the Arab population of the West Bank and Gaza Strip during the decades that followed the Six Day War. The chapter provides a descriptive overview and considers whether trends in the rate of women's labor force participation and gender-linked occupational differentiation that were observed across the world and among women in Israel also prevail among Arab women in the West Bank and Gaza Strip. More specifically, it considers whether labor force participation of Arab women in the Occupied Territories has actually increased and whether gender-based occupational differentiation has decreased during the last two decades.

Generally speaking, participation of Arab women in the market economy has been highly restricted in terms of volume, location, and jobs (Youssef 1972; Lewin-Epstein and Semyonov 1992). Women in Muslim countries face serious obstacles in joining the economically active labor force, and the complexity of this issue has been well documented in a series of chapters (Keddie 1990; Hijab 1989).

In most Arab societies, participation of women in the market economy is very low and has only recently begun to rise. This is evident in the figures published in the volumes of the *Statistical Yearbook of Labor Statistics* (International Labor Organization 1989). For example, the labor force activity rate among Algerian women is 4.4 per cent (compared to 42.4 per cent among men). In Iraq the activity rate for women is 5.8 per cent (compared to 41.6 per cent for men), and in Kuwait it reaches 13.5 per cent versus 48.9 per cent for men.

In Israel the participation rate of Arab women in the market economy has at the present time reached 15 per cent (as compared to 45 per cent among Jewish women). Indeed, this rate is extremely low, not only when compared to Jewish women but also when compared to women in many countries across the world (Semyonov 1980; Lewin-Epstein and Semyonov 1992).

TRENDS IN LABOR MARKET PARTICIPATION

Labor force participation of Arab women is strongly affected by traditional norms and values that dominate Arab societies. Therefore, one should evaluate labor force participation of Arab women within the context of the Arab value system. It has been suggested that in Arab societies 'institutional mechanisms operate effectively to isolate women from activities outside marriage and prevent them from participating in public activities which presuppose contact with the opposite sex' (Youssef 1972; 172). Consequently, the economic activity of Arab women is more likely to take place in the community of residence, where family control over community members (and especially over women) is relatively strong (Lewin-Epstein and Semyonov 1992; Semyonov and Lewin-Epstein 1994).

Most labor market activity of Arab women is likely to be limited to semi-professional white collar jobs such as teachers, nurses, and social workers, which are considered suitable for Arab women (Lewin-Epstein and Semyonov 1992). According to Boserup (1970), the concentration of women in professional and semi-professional occupations in traditional societies, such as the Arab society, can be understood in the context of the prescribed seclusion of women. In traditional societies, including Arab society, 'modern facilities such as schools and hospitals can be introduced without danger to the system of seclusion only on condition that a staff of professional women is available so that contacts between men and women belonging to different families may be avoided' (Boserup, 1970, 126).

Recent research confirms the applicability of these models to the participation of Arab women in the Israeli market economy. For example, Lewin-Epstein and Semyonov (1992) and Semyonov and Lewin-Epstein (1994) have demonstrated that Arab women are heavily overrepresented in semi-professional jobs, whereas Arab men are overrepresented in manual (unskilled and skilled) occupations.

It should be reemphasized, however, that the number of Arab women who join the cash economy in the Arab community is still rather small. Thus economically active Arab women should be viewed, in a sense, as pioneers with respect to entrance into the labor force. These women are not necessarily representative of the total population of Arab women. Indeed, the very few women who join the economically active labor force are highly selective. They are qualified in terms of education for high status professional and semi-professional jobs.

According to the literature on patterns of women's labor force participation in general and in Arab societies in particular, I expect women's labor force participation in the Occupied Territories to be low, but to increase moderately over time. I also expect considerable differences between the occupational distribution of men and that of women, with women concentrating mostly in semi-professional occupations. Finally, I expect that as their participation increases the overrepresentation of Arab women in the high-status semi-professional occupations is likely to decrease.

DATA SOURCE

Data for the present analysis were obtained from the publications of the Israel Central Bureau of Statistics. Since 1970 the Israel Central Bureau of Statistics has been conducting labor force surveys in the West Bank and Gaza Strip territories. Some of the data collected in the surveys are published annually in the *Statistical Abstracts of Israel*. From the published tables (1970 to 1990) it is possible to estimate labor force participation rates for men and women (age 15 and over) and to compare occupational distributions (in major categories) of the two gender groups at five points in time (1970, 1975, 1980, 1985, 1989).[2]

It should be noted that *Statistical Abstracts of Israel* is the only source of published information available to us on Arab women's labor force participation within the territories administered by Israel. These data cannot provide a direct examination of the effect of military occupation on women's employment opportunities. Nor can they provide any information on employment patterns in the informal economy. Also, it is likely that the figures for Arab female participation are underestimated in official statistics as is generally true for less developed economies and traditional societies (Acker 1980; Beneria 1982). Women may prefer to deny employment outside the household for either cultural or taxation reasons. This can be especially pronounced among women employed in household services or in agricultural work (where work is carried out within the family unit). Yet, these are the best data available. These data do, however, provide the basis for a descriptive overview of trends in women's employment patterns in the West Bank and Gaza Strip over the last twenty years of Israeli occupation.

FINDINGS

Participation rates for men and women at five points in time are listed in Table 8.1. The values presented in the table are somewhat surprising. Contrary to expectations based on the trends of female labor force participation that have been observed in many countries across the globe (as well as among Arab women in Israel), the participation rate among Arab women in the West Bank and Gaza Strip territories has actually declined since 1970. By contrast, participation rates for men have increased during this period. As will be argued shortly, these two opposing trends seem related to one another.

Since the 1967 Six Day War an ever increasing number of Arabs (mostly men) from the Occupied Territories have found employment in Israel. By the mid-1980s over a third of the work force of the West Bank and Gaza Strip was actually employed in Israel. Most of these workers were daily commuters and most were employed in menial, semi-skilled, and unskilled jobs (Semyonov and Lewin-Epstein, 1987). Because of the traditional norms

TRENDS IN LABOR MARKET PARTICIPATION

Table 8.1 Labor force participation of the Arab population of the West Bank and Gaza Strip 1970–1989 (age 15 and over)

	1970	1975	1980	1985	1989
Female participation (in %)	10.0	9.3	9.1	6.5	5.7
Male participation (in %)	59.6	62.5	59.9	62.3	69.3
Women's share of the economically active labor force (in %)	16.0	14.0	13.9	10.1	8.1
Size of labor force Population (in thousand)	180.8	207.0	218.5	251.2	290.33

Source: Computed from published data, Section 27, *Statistical Abstracts of Israel* (1970–1990) published by Israel Central Bureau of Statistics

and values that dominate Arab society, very few women workers could respond to employment opportunities in Israel. Consequently the overwhelming majority of these commuter-workers have been males. Women have had to stay behind in their home communities, where little economic progress and development have taken place and where employment opportunities have remained rather scarce.

Table 8.1 indicates that the size of the economically active labor force in the Occupied Territories has increased considerably between 1970 and 1989. The growth in the labor force participation may be the result of two parallel processes. The first is a mere reflection of the high level of fertility and natural growth that characterizes the Arab population. The second may reflect increased participation of workers in the formal economy and decreased rates of unemployment in the population of the territories.

The figures presented in Table 8.1 reveal that the labor participation rate among Arab women was extremely low when Israel assumed control over the West Bank and Gaza Strip. It has become even lower during the years of Israeli military occupation. In 1970 the participation rate for Arab women was 10.6 per cent; by 1989 it had declined to 5.7 per cent. Subsequently, women's share of the economically active labor force has shrunk from 16.0 per cent in 1970 to 10.1 per cent in 1989.

In Table 8.2 the occupational distributions of men and women at the same five points in time are examined. The data in Table 8.2 demonstrate rather clearly that the occupational composition of the labor force in the Occupied Territories has changed considerably during the years that followed the Six Day War. The most noticeable changes have been the decline in the size of the population engaged in agricultural activity and the sharp increase in the population employed in manual (skilled and unskilled) jobs. Examination of the data suggests that these changes have been far more pronounced for men than for women. The shift in employment location from the Occupied Territories into Israel has especially affected the occupational structures of the Arab male population. Apparently, upon joining the labor market in Israel, Arab men left agricultural jobs in the territories in order to take manual jobs in Israel.

The sharp decline in agricultural employment, though similar to the pattern of industrial transformation observed in many less developed countries, had been affected by the military occupation. That is, in 1967 the economy of the territories (especially the West Bank) was characterized by high levels of agricultural activity and high levels of unemployment (in both the West Bank and the Gaza Strip). During the years of military occupation new employment opportunities for the population of the territories became available mostly in Israel (mainly in menial-blue collar and service jobs). Consequently, many workers had left their family farms and had begun working in the Israeli labor market (Semyonov and Lewin-Epstein 1987).

Due to traditional norms and values that place constraints upon the women's labor force activity in Arab society, it was largely only men who responded to the new employment opportunities in Israel. Consequently, male participation in agricultural activity has declined more rapidly than women's participation. This process is consistent with propositions put forward by Boserup and others regarding gender-linked occupational differentiation in the process of industrial transformation in less developed economies. According to Boserup (1970) when men find jobs in the urban centers, some women stay behind and replace men in agricultural employment.

When comparing the occupational distributions of the two gender groups, it becomes apparent that women are overrepresented in the professional category and in the agricultural jobs at all five points in time. Women are, however, consistently underrepresented in the managerial, sales, service, and manual occupations. Both men and women are equally represented in jobs in the clerical category. These patterns remain quite consistent regardless of the specific year examined. The most noticeable change in the occupational structure of Arab women has come in their increased proportion in professional and semi-professional occupations. In 1970 about 15 per cent of all economically active Arab women in the Occupied Territories were employed in professional or semi-professional jobs. By 1989, one-third of all Arab women there were employed as professionals or semi-professionals. The rise in the proportion of women in the professional category seems even more dramatic when one considers the fact that the proportions of men employed in this category remained relatively stable over the years.

In the last line of Table 8.2, the values of the index of dissimilarity between the occupational distribution of men and of women are listed. The index of dissimilarity provides a summary measure of the degree of occupational differentiation between the two groups at each point in time. It

Table 8.2 The occupational distribution of the labor force population in the West Bank and Gaza Strip 1970–1989 (in percentage by gender)[1]

	1970[2]		1975		1980		1985		1989	
	Males	Females	Males	Females	Males	Females	Males	Females	Males	Females
1 Scientific professional & semi-professional	5.63	14.80	5.43	20.48	5.39	20.07	5.65	31.93	5.44	33.81
2 Managerial[2]	{3.57	{3.61	0.85	0.34	0.88	0.34	1.29	–	0.89	0.89
3 Clerical[2]			2.88	2.78	2.53	2.38	2.43	2.87	2.06	4.44
4 Sales	12.98	2.53	11.63	2.78	11.89	2.38	11.16	2.46	12.96	3.55
5 Service worker	6.94	2.17	7.65	4.17	7.81	4.42	7.90	2.87	7.28	2.67
6 Farm & agricultural workers	32.42	57.40	21.44	51.74	18.00	55.10	18.51	45.90	17.70	45.33
7 Skilled manual[2] workers	{38.46	{19.49	26.32	14.93	29.62	13.95	28.98	12.29	28.95	8.00
8 Unskilled workers[2]			23.82	2.78	23.88	1.36	24.07	1.64	24.75	1.33
N = 100%[3]	145.6	27.7	176.3	28.8	181.7	29.4	217.7	24.4	257.0	22.5
Index of dissimilarity[4]	38.57		45.36		52.28		54.13		59.78	

Source: Data were compiled from published Tables Section 27, Statistical Abstracts of Israel (1970–1990)

Notes: [1] 1989 was the last point in time for which data were available.
[2] 1970 categories were compiled to enable comparison with later years.
[3] N (in thousands).
[4] Index of dissimilarity computed for 9 occupational categories.

Table 8.3 Relative odds for men and women employed in selected occupational categories 1970–1989[1]

Occupational category	1970	1975	1980	1985	1989
Professional	2.91	4.48	4.41	7.83	8.88
Clerical	1.01[2]	0.96	0.94	1.19	2.21
Sales	0.17	0.22	0.18	0.20	0.25
Service	0.30	0.53	0.55	0.34	0.35
Manual	0.39	0.21	0.16	0.14	0.09
Agricultural	2.81	3.93	5.59	3.73	3.86

Notes: [1] See text for computation.
[2] Value based on both managerial and clerical.

indicates the percentage of either men or women who would have to change (major) occupational category if the two groups were to be equally distributed across occupations.[3]

The values of the index of dissimilarity displayed at the bottom of Table 8.2 suggest that occupational differentiation between Arab men and women in the territories has actually increased during the period of Israeli occupation. Over these years the two gender groups have grown less and less similar in their occupational structure. More specifically, in 1970 about 38 per cent of either men or women would have had to change occupations (major category) for identical occupational distributions to be reached while by 1983, almost 60 per cent of either group would have had to change occupations to obtain equal distributions.

In order to examine in a more systematic manner trends in gender-linked differentiation in specific occupational categories, a series of odds-ratios were computed. The odds-ratios capture the relative odds of women to men who belong to an occupational category (versus all other occupations). According to the coding system used here,[4] when the value of the odds-ratio exceeds unity, women have greater odds of belonging to the occupational category. When the value is smaller than unity, men have greater odds of being employed in such occupations. The value 1 indicates equal odds. The figures in Table 8.3 pertain to the following occupational categories: professional; sales; service; manual; and agriculture.

The odds ratios presented in Table 8.3 reaffirm the findings and conclusions discussed in the previous section. The small number of Arab women from the Occupied Territories who have entered the market economy hold greater odds of being employed in professional and agricultural occupations, but lower odds of being employed in service and manual jobs. As their relative size in the labor force has decreased, these women's relative odds for professional employment have increased considerably. It increased from 2.91 in 1970 to 8.88 in 1989. Females' relative odds for employment in agricultural jobs have also increased, but at a rather moderate rate. Apparently both men and women have been exiting agricultural employment, but men have been leaving such jobs in much greater numbers. At the same time, men's odds for employment in manual occupations have increased dramatically. Their relative odds for manual employment have risen from 2.56 (1/0.39) in 1970 up to 11.11 (1/0.09) in 1989. Indeed, the trend in gender-based occupational differentiation is rather clear. As the proportion of women in the economically active labor force has decreased, women have increased their relative representation in the professional and semi-professional occupations, while men have found employment in disproportional numbers in the blue-collar manual occupations (a substantial number within Israel) at an ever increasing rate.

CONCLUSIONS

The data published by the Israel Central Bureau of Statistics indicate that during the last two decades, participation rates of Arab women in the market economy of the Occupied Territories have actually declined. This trend is the opposite of trends observed in countries throughout the world. Apparently, while Arab men from the West Bank and Gaza Strip have found employment opportunities in manual labor in Israel, Arab women have remained in their communities. The traditional values of the Arab society have prevented women from searching for jobs outside their place of residence. Employment opportunities in the West Bank and Gaza Strip, however, have remained quite scarce. Consequently, the proportion of Arab women who have joined the economically active labor force has declined. The few who joined the market economy have been disproportionately employed in the professional and semi-professional occupations and to some extent in agricultural jobs. It is highly possible that these trends in labor force participation and occupational representation among women are related to the changing conditions, resulting from the Israeli military occupation in the West Bank and Gaza Strip. We still need, however, somewhat more systematic examination, using different data, of the impact of the military occupation on women's labor force processes in the West Bank and the Gaza Strip and, especially, on women's participation in the informal economy there.

NOTES

1 Many thanks to Reuben Brahm and Tammy Lerenthal for their help in data preparation.
2 1989 was the last point in time for which data were available.
3 The D-index of dissimilarity is computed according to the following equation: $D = \Sigma |P_{iw} - P_{im}|/2$ where P_i is the percent of either w(women) or m(men) in the i th occupational category. The index reported here was computed for nine major categories.

4 The odds-ratios are computed as $(f11 \cdot f22/f21 \cdot f22)$ where f is the frequency for females (1) and males (2) in occupation category (1) versus all other occupational categories (2).

BIBLIOGRAPHY

Acker, J. (1980) 'Women and stratification: a review of recent literature', *Contemporary Sociology*, 9: 25–35.
Beneria, L. (1982) (ed.) *Women and Development*, New York: Praeger.
Boserup, E. (1970) *Women's Role in Economic Development*, London: Allen (George) and Unwin.
Hijab, N. (1989) *Women Power: The Arab Debate on Women at Work*, Cambridge: Cambridge University Press.
International Labour Office (1989) *Yearbook of Labour Statistics*, Geneva: ILO Office.
Keddie, N. (1990) 'The past and present of women in the Muslim World', in *Journal of World History*, 1: 77–89.
Lewin-Epstein, N. and Semyonov, M. (1992) 'Modernization and subordination of Arab women in the Israeli labor force', in *European Sociological Review*, 8: 39–51.
Semyonov, M. (1980) 'The social context of women's labor force participation: a comparative analysis', in *American Journal of Sociology* 86: 534–550.
Semyonov, M. and Lewin-Epstein, N. (1987) *Hewers of Wood and Drawers of Water: Non-Citizen Arabs in the Israeli Labor Market*, Ithaca, N.Y: ILR Press, Cornell University.
Semyonopv, M. and Lewin-Epstein, N. (1994) 'Ethnic labor markets, gender and socioeconomic inequality: a study of Arabs in the Israeli labour force', in *The Sociological Quarterly*, 35, 1.
Youssef, N.H. (1972) 'Differential labor force participation of women in Latin America and Middle Eastern countries: the influence of family characteristics', *Social Forces*, 51: 135–153.

9

WOMEN STREET PEDDLERS
The phenomenon of *Bastat* in the Palestinian informal economy

Suha Hindiyeh-Mani, Afaf Ghazawneh, and Subhiyyeh Idris[1]

INTRODUCTION

This chapter discusses the phenomenon of women's work in *bastat*[2] – the peddling of vegetables, fruits, and other commodities on sidewalks and roadsides of Palestinian cities and refugee camps. It simultaneously involves petty commodity producers and petty traders both in non-capitalist production and in the capitalist market economy. *Bastat* work is not restricted to women; indeed many men are involved in it. However, we are concerned with the particular conditions of women *bastat* workers,[3] especially insofar as they have been shaped by the Israeli military occupation and have contributed to the changing economic status of women within Palestinian society in the Occupied Territories.

Historically, through domestic work and the provision of care to all the members of the family, Palestinian women have played a labor role in the family. This work, however, has not been considered productive. Additionally, women have participated in planting and cultivating crops and in animal husbandry on family farms; yet here too women's contributions have been considered an extension of domestic work. These familial jobs – of housekeeping and caring for children and adult males (father, brother, husband) and cultivating family land, raising poultry and livestock – are not included within calculations of either the family income or gross national income (GNI).

Both the economic pressures of survival exerted on Palestinian families by the Israeli military occupation and the political imperatives of disengagement from the Israeli economy, intensified in turn by the *intifada*, have encouraged the entry of Palestinian women into the informal market economy through *bastat* work. As a result of Israeli occupation policies imposed since 1967 (ranging from land confiscation and irrigation restrictions to competition from Israeli agricultural commodities), Palestinian male adults began to depend on different forms of wage labor, primarily those involving work in Israeli enterprises. Since before the Occupation agriculture constituted a major source of livelihood for families in the Palestinian economy, women already working in agriculture took on increased workloads. In the past, women worked alongside adult male relatives in the fields, and they continue to do so today; they also went from house to house in urban areas selling surplus goods produced on family land. Over time, as family expenses have increased and economic conditions have deteriorated as a result of the Israeli military occupation, the practice among women of selling agricultural produce has come to exceed that of raising it for subsistence. Moreover, the political aims of the *intifada*, which include Palestinian economic self-sufficiency,[4] have increased the Palestinian population's demand for locally-produced goods. As a result, the demand for family produce marketed by peasant women (the cornerstone of domestic production) in major Palestinian cities has increased.

The implications of women's entry into the informal market economy through *bastat* for their status in traditional Palestinian society are as yet unclear, or at least ambiguous. The work has largely been limited to marketing basic agricultural goods in order to contribute to the family income – both in West Bank areas and, to some extent, in the Gaza Strip, although the commodities produced and marketed there differ – so it has in some ways reinforced women's obligations, and consignment, to family life and the domestic sphere. At the same time as women have begun to market products on their own, exhibiting goods on streets and sidewalks of major cities and different markets,[5] they have been selling their goods to male merchants, particularly commission agents. In the process, women have discovered that the male merchants and salesmen to whom they have marketed products have sometimes exploited them and failed to pay them fairly for the value of their goods. Coming to terms with and negotiating these economic inequities has enabled women to learn market concepts such as profit, loss, supply and demand and to take transitional steps into the public sphere of capital.

METHODOLOGY OF STUDY

Because the work performed by women in the informal sector is not documented in the GNI and gross national production (GNP) records, we were unable to obtain statistics or basic information on women that work in this field. Hence, we collected information by interviewing women selected to represent the various types of commodities which are exhibited for sale, including fruits and vegetables, dairy and canned goods, poultry, ice cream, basic foodstuff, clothing and shoes, antiques, plastic utensils, make-up, perfumes, and textiles.

Under the prevailing conditions – strikes and the closure of areas by the Israeli occupation authorities – it was difficult for us to reach all Palestinian cities where the phenomenon of *bastat* is prevalent. Consequently, we chose East Jerusalem (13 cases), Ramallah (4 cases) in the West Bank, and Jabaliya refugee camp in the Gaza Strip (15 cases) as our study area. Jerusalem and Ramallah are geographically central to all surrounding towns and villages in the West Bank and the Jerusalem market is an open one, accessible to all who wish to market their commodities, particularly since East Jerusalem, unlike other occupied areas, has enjoyed relative calm since the beginning of the *intifada*. In the Jabaliya refugee camp, *bastat* work was a prevalent phenomenon even prior to the 1967 occupation, and large numbers of women continue to work in *bastat* in the camp.

Women were interviewed on the following subjects:

1 Marital status, educational background, place of residence, age, and, if married, husband's occupation.
2 The conditions and nature of work, including number of years of work, place of work, kind of goods sold, working hours, holidays, profits, work problems, and previous work experience.
3 Incentives for choosing this kind of work.
4 Changes occurring in working women's personal attitudes and in families as a result of their work.
5 Social attitudes towards women's work in *bastat*, and difficulties and obstacles that women face from family and society as a result of their work.

BACKGROUND OF *BASTAT* WORKERS

Twelve of the seventeen women working in Ramallah and Jerusalem are from rural areas, where they are from peasant agricultural families which have traditionally depended on farming and on rearing poultry and other livestock. The villages from which these women come – Artas, Batir, and Nahalin in the Bethlehem area; Halhoul in the Hebron area; and Dura al-Qara' in the Ramallah area – are situated on fertile lands and depend primarily on agriculture. The rest of the women in the study were from a variety of locations: two from Gaza City; two from West Bank refugee camps (Deheishah, near Bethlehem, and Qalandia, between Jerusalem and Ramallah); and one from Wadi Rahal in the Bethlehem area. Of the fifteen women who work in Jabaliya refugee camp, twelve live in the refugee camp itself; and of the three remaining, two are from Jabaliya village and one is from the nearby Tel Za'ater housing project.

The majority of women we interviewed are illiterate. They are generally middle-aged; when they were school-aged there were no schools and they were not given the opportunity to study. Even if schools had been available for them, the traditional social views of women would likely have prevented them from leaving the home to gain an education. Nonetheless, several, particularly in Jerusalem and Ramallah, have managed to gain some education, ranging from third elementary to second secondary.[6] This may be due to the fact that more schools were opened in West Bank villages and refugee camps in later years and traditional social views restricting women there have undergone a relative, if barely measurable, decline. Mass migration and the unsettled conditions to which the Gaza Strip population has been subjected because of frequent wars and control by different foreign powers have dramatically reduced the literacy rate of women in Jabaliya refugee camp.[7] The shortage of schools and/or the delay in school openings, as well as the depreciation of women's social value which have characterized traditional Palestinian society under Israeli occupation are all conditions which appear to have contributed in especially severe ways to the low level of educational attainment among the Gaza Strip women peddlers.

The majority of women in the study working in *bastat* in Jerusalem and Ramallah were either married or widowed. In general, Palestinian society views women's primary role as that of a housewife. However, the types of work in which women engage in differ between villages and towns: village women generally work in agriculture and animal husbandry; while women in towns work in other kinds of paid jobs or do not work at all. In any case, while the social value assigned to agricultural work is generally lower than the value assigned to work typically done by women in towns, especially work that does not require physical effort, the village women's marketing of agricultural products finds relative acceptance, especially if the women are married. This appears to explain the higher number of married women in our survey group.

The situation is somewhat different for single women. In rural areas the social belief that leaving home to market products in towns decreases women's opportunities for marriage means that mothers generally do not allow their daughters to do such work. Hence, the responsibility is undertaken by the mothers themselves. Moreover, both urban and rural societies view single women family members as dependent upon males for their livelihood, especially if they are uneducated. If women are educated, they might work as teachers or nurses (i.e., jobs which are more acceptable because they have a higher social value than physical work). The two unmarried women within the survey entered *bastat* work in response to specific financial needs which emerged during the *intifada*: the loss of male wage earners within the family to detention, injury or even death at the hands of the Israeli military, as well as the serious decline in work opportunities under the Occupation, has seriously threatened the economic situation of many Palestinian families.

All the women working in *bastat* in Jabaliya refugee camp are either married or widowed. This appears to be due primarily to the fact that the phenomenon of working women in the Gaza Strip is an historical one: married women who lost male support for the family and/or whose families needed to improve their financial conditions often worked there. Unmarried women, however, did not – and still do not – go out to work, due to social restrictions against sitting in public places like streets and markets, where being exposed to passersby might threaten their social and moral reputation.

In short, Palestinian women in the Israeli Occupied Territories have entered *bastat* work out of necessity which is frequently at odds with their roles in traditional Palestinian society, and which has been exacerbated by disruptions in that society caused by the Occupation. Moreover, Palestinian women have entered the informal market economy in largely unacknowledged ways which have left ambiguous social implications of this change in their economic role.

GOODS MARKETED AND WORKING CONDITIONS

The nature of the products which the women market differs according to their area of residence. The two women from Gaza City market clothing and Egyptian commodities (cheese, gum Arabic, sheets and underwear, among other goods) in Jerusalem. Likewise, the fifteen cases working in *bastat* in the market and streets of Jabaliya refugee camp sell ready-to-wear clothes, cloth and textiles, birds, poultry, sweets, ice cream, frozen juice, fruits, vegetables, perfumes, make-up, basic utensils, shoes, antiques, and used clothing. Historically, women in the Gaza Strip have long worked in the clothing and perfume trades (products imported from nearby countries, including Egypt and Jordan). Today, women from the Gaza Strip also have access to used clothes and antiques through their husbands who work inside Israel. An additional factor contributing to the prevalence of trade in non-agricultural products amongst Gazan women is the limited amount of agricultural land in the Gaza Strip.

On the other hand, Palestinian women from the West Bank predominantly market fruit, vegetables, and processed agricultural products. Women from West Bank villages in particular are generally involved in both the cultivation and the processing of agricultural goods. Of the two women workers residing in West Bank refugee camps who work in Jerusalem, one markets agricultural products, some of which she produces at home and the remainder she buys from merchants, and the other sells canned goods which she receives from a relief agency or buys from people in the refugee camp in which she resides. (This practice of reselling relief goods emerged with the deteriorating economic conditions that result from the mass migration of Palestinian families from their homes to refugee camps, in 1948 and again in 1967, difficulties which led women to look for other means of supporting themselves and their families.) The woman worker from Wadi Rahal belongs to a Bedouin community which traditionally owns no agricultural land but moves from place to place, living in tents, raising animals and marketing dairy products. With the development of Bedouin settlement programs, some members of the community began to cultivate crops such as grapes. This woman markets *malban* (sweets made out of grapes or other fruits), which she produces at home.

Working conditions for these Palestinian women are generally difficult. Women working in *bastat* in Jerusalem work long hours: a work day there may be as long as 11 to 16 hours. Some women leave for work between 4 a.m. and 5 a.m. and return home by sunset. Others, however, work an 8- to 9-hour day. In Jabaliya refugee camp and in Ramallah women work from 7 a.m. to 1 p.m., the commercial hours specified by the Unified National Leadership of the Uprising (UNLU).[8] All women in the study take the day off from the *bastat* on general strike days,[9] during which they work at home doing domestic tasks and/or agricultural work. A major impediment faced by women in this kind of work is the lack of a proper work place: these women work in the streets and in open markets where they are exposed to severe weather conditions, like extreme heat and extreme cold. Moreover, with the single exception of women working in Ramallah, who pay a specific amount of money to merchants in order to sell from sidewalk store fronts, all the women involved in the study find areas to peddle their goods through the precarious method of squatting.

If the Israeli military occupation has created conditions which have pressured Palestinian women to work in *bastat*, it has also exposed these women to a variety of obstacles and impediments. All interviewees spoke of such harassment at the hands of the Israeli authorities, including Israeli police and tax collectors who smash and confiscate their goods, and sometimes beat them.[10] All of the women also suffer financial losses as a result of Israeli military orders such as curfews[11] imposed over particular areas, which prevent them from buying goods for sale and/or from reaching their work places. Additionally, specific events during the uprising have often disrupted women's *bastat* work on particular days. For example, the events at al Aqsa Mosque in the Old City of Jerusalem on 8 October 1991[12] significantly interfered with women's work in the area; one interviewee reported that she had to stop work for ten consecutive days. In Jabaliya, women do not face the same problems from police and tax collectors as elsewhere within the Occupied Territories, but they are subject to frequent harassment from Israeli soldiers who destroy commodities, especially during Palestinian nationalist activities in the marketplace.

WOMEN'S PARTICIPATION IN *BASTAT* AND THE ECONOMIC STATUS OF PALESTINIAN WOMEN UNDER OCCUPATION

All of the women in the study with the exception of one identified financial necessity, frequently exacerbated by the Israeli military occupation, as the primary impetus for taking up work at *bastat*. In every case, their work was undertaken within the specific context of family obligations: to improve the family's financial situation by sharing with their husbands life's economic burdens; to support their families on their own in cases where husbands or older sons were absent for whatever reasons, including arrest or death; or to replace male family members working in the *bastat* dealing with police and tax officials who may be more lenient with women than with men. The single exception to this pattern among those interviewed is a woman who markets products from the family farm and dairy products from the cow that she and her husband own, who maintained that her work was not a financial necessity but rather that she worked to fill her time because she has no children.

Women in the study sample identified a number of conditions which led them to work in *bastat* rather than in other forms of work. First, the political conditions and the repressive measures practiced by the Occupation – including arrests, killings, deportation, and forced emigration – have pushed Palestinian women into work to take the role of absent men in providing for families. Since, equally important, these women do not possess qualifications and training for other jobs, they have very limited job opportunities. None of the women interviewed holds a secondary school degree, with the exception of the two from the Jerusalem sample. None of the thirty-two interviewees was formally trained in any particular skills – significantly, all believed that if they had had more specialized skills they would not have begun working in the *bastat*. This predicament is attributable to the traditional social view of women within the Palestinian society, which defines women and their future only in terms of their relation to men (i.e., as wife or mother), who alone have the opportunity for advancement. While there are large numbers of training institutions from which Palestinian men benefit, few of these institutions have been accessible to Palestinian women, especially women who are now in their sixties.

Thus women's participation in *bastat* represents the tandem effects of both their vulnerability to the economic effects of the Israeli military occupation and their secondary status within traditional Palestinian society. Work in the *bastat* is considered menial, as well as informal, labor. It was clear from the interviews that women have internalized society's negative view of their work and thus devalue themselves as Palestinian society does. This self-devaluation is what reconciles them to their work in *bastat*. This does not necessarily mean that men do not work in the *bastat*. However, for men this work is not looked down upon in the same way since a man, as the breadwinner, is obliged to take any kind of work in order to support his family.

Palestinian women's participation in the production process and in supporting the family is a widespread phenomenon in the Israeli Occupied Territories, even though their work generally remains unrecognized. Nine of the seventeen women workers in the West Bank sample (Jerusalem and

Ramallah) support their families on their own. The remainder (eight) contribute to the support of their families. These women's husbands work in farming and rearing livestock with the help of all members of the family – including the wife, who additionally works in *bastat* in the market. In Jabaliya refugee camp, six women workers support their families on their own, while nine support their families in cooperation with husbands and/or sons. An examination of the division of labor among family members clearly shows that relatively little is done by the husband, who is nevertheless perceived as the head of the household. This division is not a recent phenomenon in agricultural areas of Palestinian society. Peasant women have always worked side by side with men, farming the fields while, traditionally, it was the man who took the produce to the market or negotiated with merchants.

Thus the expansion of Palestinian women's involvement in the market is indicative of a transitional stage within Palestinian society as a whole towards the capitalist mode of production, characterized by the entry of the largest number of workers possible – both sexes as well as a variety of ages – into the capitalist market. Yet at the same time the economic benefits of Palestinian women's trade continue generally to be transferred to male members of their families. In a number of cases women's earnings are directly invested in creating or increasing employment opportunities for men, so that women are in effect working to support men's position as wage earners even as women frequently persist in describing themselves as economic dependents. Several women in the study, for example, reported that their *bastat* work had enabled them to support their sons through university and to pay the costs involved in their sons' marriage arrangements.[13] Additionally, some women were able to build houses and open commercial shops for their sons. One woman's story in particular captures the general tendency of women within the study sample to internalize the prevailing social view of themselves in terms of their husbands' social position and as their husbands' dependents. This particular woman spent her earnings on her husband's university education. Her husband was at the time of the study still unemployed, while she is now saving money so that he can buy a car in order to go out in search of work. Moreover, at the end of the day she gives what she earns to her husband because, as a man, he is the head of the household.

BASTAT COMMODITIES: STAGES IN THE PRODUCTION PROCESS

Agricultural commodities

Several forms or stages of production are apparent within the production process among women who peddle agricultural commodities. A number of women are directly involved in subsistence production, while others are directly involved in the production of simple commodities. Some women are not directly involved in agricultural production but rather purchase their goods from merchants for resale through *bastat*. In all cases, women are simultaneously involved in more than one form of production. But while the specific distribution of tasks might vary, all cases within the study reflect a process of production based on the sexual division of labor: in addition to working directly within the agricultural production unit alongside family members and/or participating in the market, in the *bastat*, women are in charge of domestic labor.

One form of agricultural production present within the sample study is that of family subsistence, through which all family members together farm and care for land plots, producing agricultural goods for family consumption and, in the case of surplus, for sale in the market. Simple subsistence production is represented by only one woman – the Bedouin woman from Wadi Rahal, in the Bethlehem area, who markets the surplus of grapes raised by her family and the *melban* which she makes at home. Additionally, several other women sell surplus agricultural products from their family subsistence farms and invest the returns in commercial goods which they market for higher profits. These women are thus involved both in subsistence production and in petty trade.

Another form of agricultural production represented in the study is simple agricultural commodity production for the market. In this case a family plants its lands continuously with seasonal, marketable products as a means of generating an income to support itself; as in subsistence production, women serve as active participants in the production process. Five of the women interviewees are involved in this form of production – planting and harvesting agricultural goods on family-owned land. A sixth woman is involved both in crop production on family land and in raising dairy cows and producing simple dairy products.

But the majority of Palestinian women involved in the marketing of agricultural products are not directly involved in production at all. These women purchase their commodities from commission agents and resell them through *bastat*. Outside the production unit (the family farm) women negotiate commercial transactions between the person marketing the household's products (e.g. *melban*, yogurt) and the client, or with retail merchants who serve as intermediaries between wholesalers and clients. For example, the woman from Artas village who markets commodities in Jerusalem became the owner and supervisor of a commercial project employing women workers who sort and clean merchandise and transfer it from place to place where they market it. In addition, she owns a small home dairy project and sells yogurt and cheese, also through *bastat*. The interviewee from Nahalin, who supports her family on her own, has also expanded her stock of merchandise in response to increased demands for her commodities.

She is now considering employing a woman to help her sell, in addition to working as a commercial intermediary. Another interviewee from Qalandiya refugee camp has likewise managed to support her family on her own, following her father's death and the arrest of the other male wage earner in her family. By working as a commercial intermediary for *bastat* commodities, she has managed to accumulate a small amount of capital with which she hopes to open a small vegetable shop. This woman originally worked in a pharmaceutical factory but left the job because the pay was not adequate for supporting her family. She has found working through *bastat* more profitable.

All interviewees market their commodities, whether homemade or bought from merchants, in the local markets by displaying them through *bastat*. They earn a small income through the circulation of commodities, perhaps accumulating a small amount of capital, and thus improve the economic conditions of their families. As mentioned above, with the exception of women working in Ramallah all the women peddle their goods on squatted ground. They need to be versatile and resourceful in their marketing strategies; the three women from Dura al-Qara' who go to nearby Ramallah to sell produce from their family farms, for example, sell both in bulk and in smaller quantities. These women explained that selling goods in small quantities brings higher profits but requires more time, and sometimes – as for example, when *intifada* activities like demonstrations, shooting, or curfews dictate – they are forced to sell produce in bulk in order to get rid of the day's merchandise quickly.

Thus while the sexual division of labor within the agricultural production unit continues to shape the commercial choices made by Palestinian women involved in *bastat*, especially in the selection of commodities, the process of steadily expanding trade remains nonetheless apparent among them, whether women sell in bulk or in small quantities. For example, the interviewee from Artas mentioned above managed to establish her dairy project with the capital that she accumulated through her *bastat* trade. The woman from Nahalin and the one from Qalandiya refugee camp are likewise planning to expand their trade. These possibilities have created sufficient incentive to the women to continue working within their current line of work, rather than considering changing jobs.

Non-agricultural goods

Fifteen women in the study market non-agricultural commodities. Additionally, two women sell a combination of agricultural and non-agricultural products.

Neither simple subsistence production nor simple commodity production is generally characteristic as a form of production amongst these women, largely because they lack access to the raw materials and the means of production necessary to produce non-agricultural goods, beyond the most basic machines. Two women own simple sewing machines and make dresses for sale, in addition to buying goods from wholesale merchants; their cases reveal the integration of subsistence economy production and simple commodity production, from producing clothes for their families to producing for the market.

Two specific cases represent the further integration of women's roles as petty producers and petty traders. One woman began her trade by producing ice cream and homemade sweets which she then peddled in the market. After she had accumulated a small amount of capital, she began purchasing similar goods directly from wholesale stores, rather than producing them herself. The second woman began her trade by knitting children's clothing and then selling it in the market. Having accumulated some capital, she then began buying ready-made clothing for resale.

The other women who peddle non-agricultural commodities are not involved in direct commodity production but, instead, obtain their goods through trade. In Jabaliya camp, most women (seven cases) buy all their goods from wholesalers or travel to nearby countries, namely Egypt and Jordan, to purchase dresses, cosmetics, perfume, and other goods to resell them through *bastat*. These women began trading in simple commodities, and expanded and developed the range of goods in which they traded as their profits increased. Three women market a combination of surplus family goods and other goods, often remnant goods that they collect themselves. The woman from Deheishah refugee camp mentioned above, for example, sells canned goods from a relief agency (both her own allotment and allotments she purchases from other camp residents). Another woman sells children's clothes and staples in Jabaliya refugee camp. She started out peddling used objects, like clothes, that her family did not need and with the capital she gained she began purchasing ready-made children's clothing which she peddled in the camp. She then began purchasing powdered milk which camp residents received from a relief agency, and sold this milk directly to wholesalers and to owners of sweatshops. As a result, she managed to accumulate further capital and expanded her trade to include the purchase of wholesale household staples. She thus became involved in trading at both retail and wholesale levels. The third woman in this group began by selling cloth remnants left over from her job as a seamstress, working out of her home, and has now developed her trade to include the purchase and resale of fabric and ready-made clothing.

In these case studies, we see Palestinian women involved in the trading of non-agricultural goods making the transition to small-scale capitalist production. A woman may simultaneously be the owner of a small-scale production unit and a worker: the woman who knits wool clothing, for example, possesses the means of production (a knitting machine) and also supplies the labor needed for production. Additionally, many women act as intermediaries, purchasing goods from merchants and reselling them through *bastat*. This includes cases in which women travel abroad to purchase goods for resale. Within the Jabaliya camp sample in particular, women have felt encouraged to develop their trade as they have increased their commercial knowledge and their understanding of supply and demand. An additional example there involves a woman who began her trade by collecting cloth remnants, linen, and sponges from local factories and selling them at a small profit. When she had accumulated a small amount of capital she opened two shops; her husband works in one store and her son in the second. One of the shops markets cloth and sponge remnants, and the other markets new fabric. Thus many of these women have increased the variety of their commodities and moved increasingly towards the sale of consumer goods which generate higher profits and allow them to earn more reliable incomes.

The marketing process in which Palestinian women selling non-agricultural goods in the Israeli Occupied Territories are engaged can be divided into several stages. These stages dramatize the transition to a capitalist market economy which these women are making, at the informal level, in response to economic and political pressures exerted by the Occupation. This is particularly clear in Jabaliya refugee camp. Most women working in the camp began their trade, out of necessity, by selling simple goods in front of or inside their own homes. As demand for their goods increased and profits accumulated these women began to expand the range and quantity of their merchandise and to look for more profitable places to sell their goods, thus entering a new stage in the marketing process. In this second stage some of the women peddlers went to UNRWA clinics, where large numbers of women regularly gather as they wait to receive medical treatment, and took up permanent squatting places there from which they now market their goods. Other women moved to the Friday market in Gaza City, and regularly marketed their commodities where large numbers of women shop. Some women were able to find storage places and to purchase and store merchandise which they resell at either retail or wholesale prices, thus entering yet another stage in the marketing process.

CONCLUSION: WOMEN AND THE PALESTINIAN INFORMAL ECONOMY

An examination of the different stages of the production process (including the organization of production, relations of production, and the marketing of products) suggests that women are currently becoming integrated into the capitalist production process which is increasingly prevalent within the Palestinian economy.[14] Significantly, while many of the women described here said that they had neither experienced personal changes nor adapted to the market world as a result of their work in the *bastat*, the same women (particularly those working in Jabaliya refugee camp) generally expressed a marked unwillingness to give up their work, even should the financial situation of their families improve. Clearly the responsibility which their work gives them has become for these women a source of satisfaction. Yet, because of its informal nature, that work remains largely unrecognized as productive work – or as a sign either of fundamental economic and social change taking place within traditional Palestinian society as a result of Israeli military occupation or of some of the limits of that change.

Palestinian women's role in the production process originates with their role in subsistence economy irrespective of the different forms of production in which they may later become involved. The extension of women's work beyond the subsistence economy (through the circulation of commodities in the market) provides for the possibility of small capital accumulation and entry into simple commodity production and even into capitalist production. That is, the fundamental position of women within subsistence economy leads to the extension of production which, in turn, develops into simple commodity production, particularly if the family owns agricultural land and/or if the family is involved in the processing of animal products like yogurt and cheese. Hence, women are involved in the entire process of simple commodity production. A woman's role within subsistence economy begins with the provision of care for her children, husband, and family. Women's domestic responsibility constitutes the social reproduction of the family's labor power. Additionally, women are involved in agricultural work to meet the basic needs of the family and further participate in the marketing of surplus goods not needed by the family.[15] Some of these women also purchase additional commodities from merchants to resell at a profit. These women are thus involved both in the production of value goods for domestic consumption and, in the case of surplus, in the production of marketable goods which carry an exchange value regulated by market prices. The accumulation of capital gained through the sale of surplus subsistence goods and the circulation of commodities in the market has allowed women to expand their trade; in some dramatic cases, women have even opened small storehouses to store their growing stocks of merchandise.

The Palestinian women in this study expressed a preference for selling their goods through the *bastat*, as this ensures that all profits accrue to the family rather than to intermediaries who buy in bulk. The marketing or selling of surplus subsistence products has led to the expansion of production; several families have been able to buy farm animals for the purpose of processing and selling products. The woman from Artas, for example, is responsible for all stages of production, from milking the cows to processing the milk to producing the milk derivatives (e.g., cheese and *lebaneh*). Moreover, through marketing processed agricultural products this woman began employing other women to clean and sort the goods in preparation for sale. Her case exemplifies the integration and dependency among the different forms of production (subsistence economy, simple commodity production) leading to capitalist commodity production.

Should this woman's work be considered free labor, though, since she neither owns the means of production (the family, and particularly the husband, maintains ownership rights) nor has the right to spend her earnings or to consider herself the owner of a small capitalist project? This question touches on some of the key ambiguities inherent in women's economic status in traditional society, left intact by Palestinian women's participation in the informal economy. The possibility of developing this woman's project into an explicitly capitalist one exists – yet, at the same time, in this context women remain dependent on their menfolk within a patriarchal society[16] even as the Palestinian economy remains, in turn, dependent on the Israeli economy. Capitalist patriarchy directly benefits from subsistence production, as the payment of low wages to male labor power does not threaten the male laborer's illusion that he earns enough to meet the needs of the family, since women are expected to provide for the basic needs of the family free of charge through their involvement in subsistence production.

During the first three years of the *intifada*, the marketing of local Palestinian goods has expanded, increasing the role of merchant capital in the Palestinian economy. This increase has led to greater involvement of women in different economic roles.[17] Yet merchant capital appears to reinforce the domination of the capitalist mode of production in Palestinian society, a situation which is characteristic of Third World countries and is a result of economic dependency on the capitalist industrial world.

How, then, shall we define the emergent role of Palestinian women in relation to merchant capital?[18] Several cases depict a distinctive relationship between women's economic role in extra-domestic activities and merchant capital, on which other significant power relations turn. Within the case studies, several women buy goods with merchant capital (through merchants in the market) and sell their merchandise through *bastat* in various locations. The woman in Jabaliya refugee camp who was able to rent two shops through the accumulation of simple capital reveals the important role of merchant capital as an agent of growth and change within the Palestinian economy. Yet this case and its representation of the role of merchant capital raise another analytical problem: one of the shops is run by the woman's husband and the second by her son. How shall we categorize this woman? Does she represent merchant capital and, through employing workers, come to represent capital itself? Or, as in the earlier case, does she represent free wage labor, since she does not own the means of production or have the right to spend her profits?

The devaluation of women's involvement in capitalist production results from the separation between subsistence production and capitalist social production which, as Bennholdt-Thomsen (1981: 41) maintains, is a basic contradiction within the capitalist mode of production. Women's work in *bastat* involves several integrated forms of productive labor: subsistence production (from housekeeping to care of family members to the harvesting and processing of agricultural products for home consumption); simple commodity production; petty trade. Nonetheless women's work remains largely unpaid and women's economic productivity remains largely unrecognized. Women do not own the means of production insofar as ownership of any means of production available to her remains in the hands of the male members of the family. And although what women do clearly leads to increasing the family's income, women's self-perception continues to be that they are dependent on men. At the same time, men continue to consider women unemployed, viewing any work that they perform as no more than an extension of their domestic (unpaid) labor.

Thus Palestinian women's entry through *bastat* work into the informal market economy has marked a potential shift in the economic gender roles (the relationship between men and women) within Palestinian society and in the relationship between the Palestinian and the Israeli economies – yet, at the same time, it has also reiterated some of the key ambiguities surrounding autonomy and dependence in both relationships and, in turn, characterizing the Israeli military occupation itself.

NOTES

1 We would like to thank Salim Tamari for his careful reading of the Arabic and English versions of this chapter and for his helpful, detailed comments. Thanks also to Shary Lapp for editing the English version and for her invaluable insights.

2 *Bastat*, the plural of *bastah*, is a colloquial term derived from the Arabic root b.s.t which means 'to spread'. The term is used to refer to the 'spread' or display of goods for sale on sidewalks and roadsides. In colloquial conversation the term refers to the act of peddling and to the places where such peddling takes place. Hundreds of women in the West Bank and the Gaza Strip sell vegetables, fruit and other goods through *bastat*.

3 Generally, men are more likely to work in more established *bastat* (i.e. permanently located stalls which can reach the size of small shops) or to use movable carts from which they sell their goods. Women, on the other hand, more often display their goods directly, along city and camp streets.

4 Economic self-sufficiency has been a fundamental objective of the Palestinian *intifada*. This goal represents an active rejection of the attempt to subjugate the Palestinian economy in the occupied West Bank and Gaza Strip to the interests of the Israeli economy which has informed Israeli policy since the Occupation began in 1967. Developing the production of local goods, particularly agricultural products, has been a central strategy in the Palestinian effort to achieve self-sufficiency.

5 This does not, however, mean that only women participate in this form of work. Nor does it imply that *bastat* is a new form of marketing.

6 The school system in the West Bank and Gaza Strip is based on the three-tiered cycle, including a six-year elementary program for children from 12 years of age, a three-year program for students from 12 to 15 years of age, and a three-year secondary program for students from 15 to 18 years of age.

7 Historically education in the Gaza Strip, which fell under Egyptian jurisdiction prior to the 1967 Israeli occupation, has been less developed than in the West Bank, as Egypt did very little to develop education there.
8 The UNLU is an underground, popularly-based leadership which emerged to coordinate uprising activities and to provide the *intifada* with a cohesive direction. In accordance with calls from the UNLU, Palestinians have continuously observed a daily half-day commercial strike for over three years as a form of ongoing protest against the Israeli occupation.
9 Because the right of Palestinians to strike is not recognized by the Israeli occupation authorities, commercial and general strikes (during which all activities in the Occupied Territories, including business, education and transportation, are brought to a complete halt) have played an important role in the *intifada*. In addition to the ongoing half-day strike, general strikes called by the UNLU to mark special occasions are observed on average four times per month. For details, see Jerusalem Media and Communication Center (JMCC) 1989.
10 During the *intifada*, long-standing Palestinian resentment of Israeli taxation policies in the Occupied Territories developed into an organized campaign of tax resistance. Refusal to pay taxes was viewed as an essential aspect of the *intifada*'s aim of disengagement from the occupying power. The Israeli authorities responded with a concerted tax collection campaign involving the imposition of exorbitant taxes and widespread intimidation to force tax payment (confiscation of identity cards and business papers, mass arrests, vandalism and/or expropriation of property, and denial of travel permits, car licenses and other essential documents). For details, see JMCC 1989.
11 Throughout its occupation of the West Bank and the Gaza Strip, and particularly during the *intifada*, Israel has used curfews as a means of controlling and punishing the Palestinian population, in violation of international law. According to the Jerusalem Media and Communication Center (JMCC), during the first three years of the uprising, Israel imposed round-the-clock curfews on Palestinian communities on more than 7,800 different occasions. Many of these curfews remained in force for a week or longer. On average, every Palestinian in the Occupied Territories spent 69 days during the first three years of the *intifada* under curfew. In bringing to a complete halt all activities of normal civil life in targeted communities, curfews have had a devastating effect on all aspects of the Palestinian economy. For details see JMCC 1991: 1, 45–48, 80–94.
12 On 8 October 1990, Israeli troops opened fire on Palestinians gathered in the al-Aqsa compound (the Muslim holy site comprised of al-Aqsa mosque and the Dome of the Rock), killing 17 Palestinians and injuring over 150 others.
13 In addition to paying the expenses for wedding celebrations, the groom is expected to purchase gold for the bride and to build or rent, as well as to furnish, a household for the couple.
14 This analysis is intended as a preliminary contribution to discussion of women's role in the various aspects of the production process, which will require more detailed and extensive studies.
15 This pattern has been widely observed throughout the Third World.
16 This problem is rather universal and not restricted to Palestinian society.
17 For example, large numbers of women entered wage labor in Palestinian factories, while other women were involved in processing and selling domestic products at home or at *bastat* or to merchants.
18 Within this discussion merchant capital is defined as money circulating in the

market among producers, intermediaries and consumers (rather than in production). The function of merchant capital is to organize the circulation of commodities within this economic matrix.

BIBLIOGRAPHY

Bennholdt-Thomsen, V. (1981) 'Subsistence production and extended reproduction', in K. Young, C. Wolkowitz and R. McCullagh (eds) *Of Marriage and the Market: Women's Subordination Internationally and its Lessons*, London: Routledge and Kegan Paul, 41–54.

Jerusalem Media and Communication Center (1989) *The Intifada: An Overview (The First Two Years)*, Jerusalem, December.

—— (1991) *No Exit: Israel's Curfew Policy in the Occupied Palestinian Territories*, Jerusalem.

10
ENVIRONMENTAL PROBLEMS AFFECTING PALESTINIAN WOMEN UNDER OCCUPATION
Karen Assaf

> We render special tribute to that brave Palestinian woman, guardian of sustenance and life, keeper of our people's perennial flame.
> (Declaration of Palestinian Independence II, November 15, 1988)

The war of 1948 shaped the demographic and environmental future of Palestine almost as palpably as a natural upheaval or disaster. Seven hundred and fifty thousand Palestinian refugees were displaced from their homes in 1948 with two-thirds of them settling in refugee camps in what is currently known as the Occupied Palestinian Territories: the West Bank and Gaza Strip. In turn, the war of 1967, as a result of which Israel occupied the West Bank and Gaza Strip, created even more environmental upheaval when another one-half million Palestinians fled these areas to neighboring Arab countries, 220,000 of them for a second time. It is estimated that three-fourths of the Palestinian population has been displaced more than once. In environmental terms the burden of coping with the last fifty years' cycles of occupation, expulsion and expansion by the Israeli state has fallen disproportionately upon Palestinian women.

Women in the Palestinian Arab culture have, for example, had to carry a disproportionate share of the burden of moving and adjusting to a new environment every time an Israeli military attack resulted in an exodus of Palestinians from their homeland. In 1948 Palestinians left well-established cities such as Safad, Tiberias, Acre (Akka) and Haifa in northern Palestine; Lod, Ramleh and Jerusalem in the center of Palestine, and Yaffa (Jaffa), Askalan and Ashdod in the south – thousands of women had to take charge of setting up homes in makeshift refugee camps.

The refugee camps set up by the United Nations near major Palestinian cities, most of which have evolved into slums of small huts with metal roofs, seem to have become a permanent feature on the landscape of the Occupied Territories. A traveler now going from north to south through the West Bank and into the Gaza Strip will find nineteen refugee camps in the West Bank and eight in the Gaza Strip. The ratio of refugees to the general population in the West Bank is over 38 per cent, while the ratio of the refugee population of the total population of the Gaza Strip is nearing 85 per cent. In the West Bank 26 per cent of all registered refugees live in camps, while in the Gaza Strip the figure is over 55 per cent (UNRWA 1992). It should be noted that five of the eight refugee camps in the Gaza Strip have populations larger than those of the majority of towns in the West Bank.

Palestinian society in the West Bank portion of the Occupied Territories is essentially an agrarian society, composed of several hundred small villages and refugee camps with only five urban areas: Jerusalem/Ramallah/El-Bireh in the center; Nablus, Jenin and Tulkarem in the north; Jericho in the Jordan Valley and Hebron in the South of the West Bank; and Gaza City and Khan Younes/Rafah in the Gaza Strip. Population density in the Gaza Strip camps averages 3,590 persons per square kilometer (Abu Safieh 1992) and reaches up to 100,000 persons/square km in Gaza City Beach Camp. These figures should be compared to comparable ones for the West Bank, where the overall density ratio is 367 persons per square kilometer, reaching 21,000 per square kilometer in Jerusalem and merely 12,000 persons/sq km in Nablus (Kittaneh and Hassan 1993). In Israel, in contrast, the overall population density is only 220 persons/square km (Heiberg and Ovensen 1993).

The task of coping with the pressures of daily life exerted by these difficult, crowded conditions is largely fulfilled, in this traditional society, by women. Palestinian women living in the Occupied Territories have also had to cope with the disruptions to the agricultural way of life and the inequities in distribution and access to natural resources which have followed the Israeli occupation.

Further restrictions imposed by the Israeli military occupation have caused overuse and misuse of natural resources and damaged the health and standard of living of Palestinians. Israeli governmental policy under occupation has been to encourage the establishment of Israeli settlements on occupied and confiscated Palestinian lands and the exploitation of Palestinian natural resources, especially water. The occupying government has also placed restrictions on land use by Palestinians and has limited Palestinian authorities' ability to control and regulate public services, major infrastructure development, and the building of homes. Israel's combined effort to reduce the natural resources available to Palestinians in the Occupied Territories while preventing Palestinian environmental autonomy has, inevitably, caused serious environmental problems, many of them with particular implications for women in the Occupied Territories: lack of water for Palestinian agricultural and domestic consumption; lack of land for Palestinian residential expansion and for Palestinian agricultural land and pastures; water scarcity and pollution; lack of sewage and solid waste control; misuse of pesticides, herbicides and other agro-chemicals; noise pollution and air pollution, mainly from quarries and neighboring Israeli factories and farms.

At the outset, it must be emphasized that the environmental problems that are currently being focused on in the international arena are not the same as the environmental problems found in the Palestinian Occupied Territories. The goal of the 1992 United Nations Conference on Environment and Development – the Earth Summit – aimed to lay the foundation for a global effort to ensure a more equitable and sustainable future for the planet and to make the environment an issue central to policy-making in every sector of economic life. The United Nations and the developing countries have focused on global challenges such as climate change, marine pollution and desertification – but these are not priority issues in the Occupied Territories, where the more immediate environmental problems are being exacerbated by the Israeli occupation. Other broadly defined issues of global concern, however, such as human resources, deforestation and sustainable agriculture, do reflect the environmental priorities of the Occupied Territories.

The quantitative restraints of the world environment have finally been recognized at an international level, and a definite link has now been established between 'environment' and 'development'. A declaration made at the Earth Summit confirming the sovereign right of states to exploit their own resources (without damage to others), the importance of international cooperation in eradicating poverty, and the role of women in sustainable development is of special interest to Palestinians in the Occupied Territories. An international strategy was also presented there that aims to achieve sustainable management of resources, building on the positive links between development and the environment and breaking the negative links between economic growth and environmental deterioration.

Thus the Occupied Territories are on the threshold of a promising development opportunity, as plans for the future can now be made on the basis of positive and negative experiences of others. Yet the Israeli state has in effect suppressed all forms of development in the Occupied Territories since before 1967, since the mere existence of the hostile and militarily strong State of Israel on the borders of the West Bank and Gaza Strip has caused insecurity and thus lack of normal development there. Moreover, since 1967 the Israeli military occupation and military rule of the West Bank and Gaza Strip have more directly restricted an openly and freely developing economy there, even though Israel claims that its presence has promoted development (Heiberg and Ovensen 1993). These 'economic' and 'environmental' conflicts have been the source of enormous stress for the Palestinian women who have had to cope with them. At the same time, the pressures to achieve local autonomy and self-sufficiency which these conflicts have exerted have, especially since the advent of the *intifada*, mobilized Palestinian women politically and given them new force in the public arena.

THE PALESTINIAN WOMEN OF THE OCCUPIED TERRITORIES

Like women everywhere in the world, Palestinian women vary in social standing, occupation, educational achievement and ideological persuasion. Yet all Palestinian women in the Occupied Territories are affected in one way or another by the Israeli occupation. Even more, 'in a crisis – and especially under occupation – it is the woman who pays the price' (Khass, in Lipman 1988: 148). Palestinian women carry the full responsibility of the home and its environs – even more so than in many traditional cultures, because the political situation in the Occupied Territories has placed such extreme pressures on the men who are frequently out of work, overworked, out of the country, in prison, or even dead.

It must be noted that it is much more difficult to draw a clear line between the effects of the environment on women in particular and on society as a whole than it is to draw such gendered distinctions for effects of social and political conditions. What must be determined is the effect of specific environmental factors, especially place of residence and lifestyle, on each group of Palestinian women. Therefore, for the purpose of this chapter, Palestinian women in the Occupied Territories will be classified according to their living environment, specifically as follows.

Refugee camp women

Women in Palestinian refugee camps under Israeli occupation must cope in both the cold winter and the hot summer with heavy pollution and running sewage. Women have to suffer more than men in this harsh, overcrowded and polluted environment because men generally go to work outside the camps for long hours or even for days at a time. The suffering of women in refugee camps is further, and unequally, exacerbated by their lack of opportunities to achieve change. Those Palestinian women who complete the preparatory level of education provided by UNRWA (the United Nations Relief Work Agency for Palestinians) find their chances for further educational training limited, even though many of them continue through high school and college levels. Moreover, while many Palestinian women activists from the refugee camps in the Occupied Territories are well known Palestinian leaders, many of them have also had to cope with stressful conditions in Israeli jails; and many have suffered miscarriages as a result of exposure to toxic gas thrown by Israeli soldiers.

Village women

Although the Palestinian refugee camps are basically urban in nature and thus women there have maintained cultural contacts with nearby cities, the

Palestinian women who live in villages have in some ways more freedom to move about. Responsibility for the environment around a Palestinian village involves taking care of more things than just the 'house'. Reducing pollution in the villages, for example, means collecting leftover agricultural produce and taking it out of the home; helping in the fields and orchards by weeding and harvesting; storing agricultural produce and materials; securing sufficient water for family use; keeping sewage and solid waste away from the living area – and women's work is essential to the tasks of maintaining a garden around the house, bringing water for the growing plants, and picking olives, fruits and vegetables, and thyme and other wild mountain spices with the children. Village women's involvement outside the micro-environment of the home is further enlarged by the fact that managing farms and selling agricultural produce in the cities nearby are other acceptable activities for them; and even riding horses and donkeys is not uncommon for women in villages.

Urban women

Palestinian cities, like cities elsewhere in the world, provide more efficient public services and have a better developed infrastructure than the villages and refugee camps. Wastewater and solid waste disposal and water supply are regulated to some extent. Electricity is provided on a 24-hour basis. Transportation and general communications are more readily available. As a result, Palestinian cities provide more opportunities for women, especially the young, who can complete high school and receive vocational and higher education in their vicinity.

THE EFFECTS OF THE ISRAELI OCCUPATION ON THE PALESTINIAN ENVIRONMENT

Lack of Palestinian residential land: the calamity of refugee camp life

According to recent reports, there are 2.2 million Palestinians living in the Occupied Territories. Sixty-five per cent of the Palestinian population lives in the West Bank (50 per cent in the north, 33 per cent in the center, and 17 per cent in the south) and the remaining 35 per cent resides in the Gaza Strip (48 per cent in the north, 19 per cent in the center and 33 per cent in the south). Of the West Bank population, 42 per cent is urban, 32 per cent is rural and 26 per cent lives in refugee camps (Shqair 1993, Heiberg and Ovensen 1993, Haddad 1992, and Assaf and Assaf 1985).

The Gaza Strip has one of the highest population densities in the world, with an annual growth rate of up to 4.5 per cent. The population there is expected to reach 1,000,000 by the year 2000. Agricultural land is being encroached upon for residential purposes. Yet as of 1991, more than 16 per cent of the land area in the Gaza Strip was used for Palestinian dwelling areas and only 45 per cent for agriculture, while 10 per cent was being used for Israeli settlements and the remaining 28 per cent, being forests and sand dunes, is considered 'governmental', and thus claimed by Israel (Abu Safieh 1992). Meanwhile, over 39 per cent of the Palestinian population of the Gaza Strip lives in refugee camps.

In the West Bank, 6 per cent of the land is used for Palestinian dwelling areas, 32 per cent for agriculture, 32 per cent for forests, pastures and grazing, and 30 per cent is classified as 'unusable'. Israel controls directly over 53 per cent of this area of the West Bank either as Israeli settlements, military areas, 'state of governmental' land, abandoned property, or simply by outright land confiscation (Haddad 1992). The Israeli occupation authorities and specific Israeli military orders and land-use master plans restrict Arab use of all Israeli controlled areas as well as lands not specifically designated within Palestinian municipal or village boundaries.

Palestinians' lack of authority over their own land, following Israeli occupation, has directly contributed to numerous environmental hazards with particular implications for women. There is a general lack of housing for all Palestinians currently living in both the West Bank and Gaza Strip. A conservative estimate is that at least 100,000 housing units would be needed within the next few years to accommodate the current population (Assaf 1989). Crowding in the tin-roofed refugee camps, which has become overwhelming in the Gaza Strip, especially around Gaza City, makes it difficult if not impossible for Palestinian women of the refugee camps to preserve sanitation within and around the home. Because Palestinian families may not be granted building licenses even if they own land, because of the strict building regulations imposed by military rules, women are forced to live in and manage homes which house several generations of a single family. Family tensions frequently increase when over eight people live together in a single room, and daughters are frequently encouraged to marry young in order to make space for the other members of the family.

Water for human use and water pollution: contamination and trips to the well

The issue of Palestinian water rights is a priority item both on the local and international scene, and has already received considerable attention from observers interested in the Middle East in particular and the developing world in general. A few points which highlight the role of the Israeli military occupation in exacerbating these environmental hardships and inequities need to be emphasized here.

Of the total ground water available in the West Bank, 86 per cent is utilized by Israel and by the Israeli settlements built within the Occupied Territories, with only 14 per cent remaining for the use of Palestinians

(Al-Khatib and Assaf 1993, Assaf and Assaf 1986, 1985). In most urban centers within the Occupied Territories nearly 100 per cent of the houses have piped water; but there is not always water in the pipes, especially during the summer. The negative pressures resulting from this situation create an ideal situation for water pollution within the piping network. Over 42 per cent of rural West Bank Palestinian communities (about 36 per cent of the Palestinian population) do not have a piped water supply (Barghouthi and Daibes 1993), and while most refugee camps have piped water due to their proximity to urban areas, the supply is subjected to frequent cut-offs. In fact, 8 per cent of all residential areas that do have a piped water system only have water in the pipes for two or three days a week (Personal communication, Palestinian Hydrology Group 1993).

Environmental hardships and hazards resulting from the Palestinians' lack of authority over their own water resources place special pressures on women. Especially in villages, women carry the responsibility of obtaining water for home use when piped water is not available, and many Palestinian village and refugee camp women must still walk long distances to shallow wells and small springs in order to fill buckets and large tins with water for home use. Since the water in the village wells and springs is not usually protected from or tested for contamination, many Palestinians are drinking polluted water. Moreover, during the *intifada* many reports were made of efforts by Israeli soldiers to contaminate water tanks in urban areas of the Occupied Territories with urine and excreta. Water tanks on the roofs of Palestinian homes were also repeatedly shot at by Israeli soldiers and settlers so that Palestinians would not have household water storage during water supply cut-offs and, especially, during curfews. The problem of obtaining adequate water supplies in urban, village, and refugee camp areas makes it difficult, in turn, to keep the household and members of the family clean and healthy.

Sewage and solid waste: the Palestinian urban crisis

Sewage disposal and treatment is becoming an important Palestinian national problem, but it needs to be handled at the local level. And this again points both to the absence of functioning local Palestinian authorities who can plan and coordinate this major infrastructure undertaking and to the added burden placed on women by environmental problems in the Occupied Territories caused or aggravated by the Israeli military occupation.

By any standard, the Occupied Territories are lagging when it comes to public services handling sewage and solid wastes. Currently the West Bank and Gaza Strip generate an estimated 47 to 56 million cubic meters (MCM) of waste water with a projected estimate of 96–122 MCM by the year 2000 and 223–283 MCM by 2020 (Al-Khatib and Assaf 1993). About 39 per cent of the houses in the Gaza Strip are currently connected to sewage systems, while the rest use septic tanks, unprotected boreholes, or open channels (Abu Safieh 1992). In the West Bank, an estimated 50 to 60 per cent of urban waste water is collected in pipes, within residential areas only, and then left to flow out in open channels (Haddad 1992). In the Gaza Strip, untreated sewage is dumped directly into the sea. Open channels of raw sewage actually run through the streets of residential areas in many cities, villages and of course refugee camps. In addition, the sewage rivers in the *wadis* and leaking household cesspools are probably draining into the water aquifers and will in the future foul drinking water in the Occupied Territories. And these problems are caused not only by Palestinian sewage but also by sewage flowing from Israel and from Israeli settlements in *wadis* near Palestinian cities and villages and over Palestinian agricultural lands.

Sewage disposal and treatment has been systematically neglected since the days of the British Mandate that ended in 1948. All types of development in the Occupied Territories have been more directly hindered by the Israeli occupation. Whatever sewage facilities that do exist there either work inefficiently or not at all due to continuing overpopulation and thus overload of existing facilities, and of course to lack of proper and regular maintenance over these years. This again points to the absence of local Palestinian authority over such major infrastructure projects. Properly treated sewage water also needs to be added to the potential water supply budget for agricultural purposes in the future.

With regard to solid wastes, 'garbage out' is not 'garbage gone', especially when the Israelis use the Occupied Territories as a dumping site (*Al Quds* 1993). In urban, village and refugee environments adequate and proper solid waste collection and disposal simply does not exist – and, again, there is no functioning local Palestinian authority to regulate, monitor and guide municipalities and communities in proper management. Both sewage and solid waste environmental hazards are, further, precursors to another environmentally induced health hazard, insect and pest hazards.

Most Palestinian men leave the house to work, leaving the garbage disposal responsibilities to the overworked Palestinian women. Women in cities, villages and refugee camps in the Occupied Territories have direct and almost sole responsibility for keeping the immediate environment sanitary, with only minimal help from other people or organized authorities. Palestinian urban areas, especially surrounding the vegetable markets (*souqs*), are almost invariably strewn with garbage. Women, who are primarily responsible for going to the market and shopping, are forced to pick their way through piles of rotted fruits and vegetables and other waste. Women also find it difficult to dispose of home refuse in city refugee camps and village areas. Where disposal facilities are present, they are few and far apart.

Lack of Palestinian agricultural land, forests and pastures and lack of water resources for agricultural use

Since the West Bank and Gaza Strip are basically agricultural, water availability is fundamental to any plans for increased production. But land in the Palestinian Occupied Territories which may be used for agriculture and grazing is limited, and land use there is currently regulated by the Israeli authorities. Additionally, virtually all water resources in the West Bank and Gaza Strip are ultimately controlled by the Israeli occupation authorities, and there is little likelihood that the Palestinian farmer will in the foreseeable future be able to increase the supply of water available to him or her.

In 1966, a total of 292,217 hectares of land were being cultivated in the West Bank. This cultivated area had decreased to 167,190 hectares by 1981 with a slight rise to 174,300 hectares by 1992; meaning a 40 per cent decrease during the past twenty-six years of Israeli occupation. In the Gaza Strip, 16,300 hectares were cultivated in 1967/1968 and 16,500 hectares in 1992, showing essentially no development in the agricultural sector there since the beginning of the Israeli occupation. Both the decrease in the cultivation of agricultural land in the West Bank and the lack of agricultural expansion in the Gaza Strip – despite the rise in the Palestinian population during this period – can be ascribed to Israeli controls over the land and water of the Occupied Territories (Abu Safieh 1992, Assaf and Assaf 1985).

Israeli land confiscation and restrictions on water use for agriculture are not the only obstacles to agricultural development in the Occupied Territories caused by the occupation. In addition, Israeli military orders require every Palestinian to obtain licensing to plant trees, even on his/her own land. In the Gaza Strip, the planting of even a single productive tree (such as an orange tree) requires official approval from the Israeli authorities. These restrictions have had a negative effect on the amount of forested land in the Occupied Territories; much of the forest was cut down during Ottoman Rule for railroads and never replaced. The total area of forests in the West Bank reaches only 4.5 per cent of the total land area, with 80 per cent of these forests in the northern Nablus and Jenin areas (Shqair 1992). The desertification and soil erosion have not yet been measured but must almost inevitably be substantial.

Moreover, it is common Israeli practice to cut down trees on Palestinian lands, under the pretext of security measures or of the government's need for the land for road infrastructure development (usually roads to new Israeli settlements). Since the beginning of the *intifada*, a total of 154,251 fruit producing trees have been destroyed on Palestinian lands. In addition, 40,846 hectares of Palestinian land have been confiscated by Israel during this six-year period alone (Palestine Human Rights Information Center 1993; Land Research Committee 1993).

Israeli limitations on areas of land to be cultivated have caused overgrazing and overuse of land in the Occupied Territories, leading to desertification and soil degradation there and prompting rural Palestinians to immigrate to cities as agriculture has become less and less profitable. Village women have therefore had to adapt to new and difficult, more crowded and cramped conditions. Reductions in family income caused by the inadequacy of allotments of water available for irrigation of cultivated land in the Occupied Territories have followed losses in Palestinian crop yields (which are, in addition, usually inferior to Israeli products produced under irrigation and thus fetch lower economic returns) and placed additional pressure to make ends meet on village women. Village farm women living in dry-farmed areas (i.e., areas without irrigation), which represent over 90 per cent of cultivated Palestinian land, have meager resources and a harsh way of life. Basic needs and necessities that are taken for granted by urban women and even most refugee women (such as electricity and piped water) are not available to village women.

Lack of pesticide control: can we blame the farmers?

Pesticides are generally used freely and abundantly by Palestinian farmers, without regulation or control, in the production of fruits and vegetables. Israel has not sought to reduce the contamination of Palestinian produce by pesticides or herbicides; Palestinian products cannot be sold in Israel in any case. Moreover, it is Israelis who sell the agrochemicals used in the Occupied Territories. Pesticide usage in the Occupied Territories is in fact a result of proximity to Israel and of the Occupation, which has limited the growth of the traditional Palestinian agriculture by restricting and/or confiscating land areas to be used for agriculture and pastures and thus persuading Palestinian farmers to use pesticides extensively in order to increase output.

Pesticides and herbicides have posed from the earliest incidence of their usage one of the most serious environmental problems in the Occupied Territories, as in the world at large. In neighboring Israel, more than 700 preparations are licensed for use in agriculture and in home gardens – including insecticides to protect crops against a variety of insects; fungicides against molds and fungi; herbicides to control unwanted vegetation; defoliants and bactericides against crop disease; rodent poisons; bird repellents and poisons; and post-harvest preparations to preserve fruits and vegetables from spoilage. Many of these agrochemicals are extremely dangerous when ingested or inhaled or absorbed through the skin, even in small quantities. Some of them are carcinogenic (cancer-producing or enhancing), some are mutagenic (can affect the genetic code and cause mutations), while others are teretogenic (have an adverse affect on embryos) (Safi and El-Nahhal 1992). Some of these chemicals are classed in what environmentalists call 'the dirty dozen' that are no longer allowed in the USA and in most of Europe, and are severely limited and strictly monitored in other countries. These preparations and materials are found in Israel and are thus found and used extensively in the Occupied Territories; and many Palestinians see using them as a sign of advancement and development in agricultural practices. Israel has thus introduced these chemicals into widespread use among Palestinian farmers in a manner which is unsuited to the degree of awareness and lack of proper regulation in the Occupied Territories.

Agrochemical poisoning can occur not only as the result of use of pesticides and herbicides by Palestinian farmers on Palestinian lands but also as a result of air drift from, for example, aerial spray organophosphate-based pesticides used for the spraying of Israeli cotton fields. These chemicals can cause side effects in both Israeli and Palestinians populations up to two kilometers away. Conditions frequently not recognized as direct pesticide poisoning include gastroenteritis, colds, eye inflammations, rheumatic cramps, insomnia, nervousness, nausea, dizziness, diarrhea and headaches. These pesticide preparations and materials are also present on many of the fresh fruit and vegetables that are eaten in the Occupied Territories, and Palestinians are frequently affected by these agrochemicals more than Israelis since, due to the economic situation, many Palestinian families cannot afford to buy meat and therefore their diets are vegetable-based. Also, agrochemicals enter the public water supply in the Occupied Territories both through the air to Lake Tiberias, which feeds into the Israeli water network and reaches some of the Palestinian population, and by leaching into the aquifers from the soil of agricultural areas on the surface.

Palestinian women are especially vulnerable to pesticide poisoning because they are exposed not only in the workplace, as agricultural workers, but also in the home, as they pick and wash food for consumption by their family and as they handle and wash the clothes of the male workers in the family who handle such substances. Because Palestinian farmers do not fully understand how to use pesticides, they frequently do not recognize all the dangers involved in simple contact with these materials. Instructions in Arabic are rarely printed on cans, bottles or jars of pesticide for sale in the Occupied Territories; even English instructions are generally overprinted in Hebrew. Storage and applications procedures are hardly ever specified, and disposal methods for used cans, bottles or sacks are rarely provided. Testing for pesticide residues is not performed in the Occupied Territories in a regular, standardized manner. However, surveys and spot checks which have been performed show that 37 per cent of the farm produce tested had pesticide residues above World Health Organization (WHO) acceptable limits (Abu El-Haj 1992). Cucumbers seem to be the most dangerous vegetable at this time; however, grapes, strawberries, peaches, and even watermelons have also been found to carry pesticide residues, even after being washed.

International norms are not being observed in the Occupied Territories with regard to pesticide and herbicide availability and use. Pesticide and herbicide misuse – to the ultimate detriment of Palestinian farmers and their families – is clearly one of the most direct effects of the Israeli occupation, of Israel's control of the economic base of the Occupied Territories and Israeli policies of promoting technology without proper and fully understood agricultural extension services. Virtually all Palestinian farmers and their families have suffered as a result of these environmental hazards. The greatest threat is posed to the women who must provide safe, uncontaminated produce for their families. The serious difficulties which they face in this task are compounded by the risks of agrochemical poisoning which women run as agricultural workers and in their other domestic duties.

Moreover, the underdevelopment of industries, coupled with the Occupation, has resulted in noise and air pollution in the Occupied Territories which poses potential environmental problems and occupational health hazards. Women's consignment to the confines of the neighborhood, the market and the village puts them at special risk here. Due to the lack of proper expansion and the consequent cramping within Palestinian cities under occupation, urban areas within the Palestinian Occupied Territories, especially around vegetable markets, or *souqs*, in the center of town where street peddlers cram the sidewalks, are often excruciatingly noisy during the daytime. During the *intifada*, the sound bombs which have been used by Israeli soldiers to disperse people have in some cases caused temporary and even permanent deafness as well as actual skin ruptures when sound bombs have hit close to the body. And particles of matter from poorly located stone quarries in the West Bank have caused various lung diseases in neighboring refugee camps, villages and towns (Fakhouri 1993).

POTENTIAL CHANGE

Since its inception at the end of 1987, the Palestinian *intifada* has sparked substantial growth in organized women's activities. Depending on their political affiliations and interests, Palestinian women have formed women's committees in the camps, the villages, and the cities. Equally important, in order to increase their self-sufficiency they have also formed special income generating organizations, including small food canning, bottling and packing factories for pickles, fruit, preserves and spices. Health relief committees, health aid organizations, mother care, food and nutrition centers and political action committees have all been formed by Palestinian women.

Thus, one may say that the Israeli occupation has increased Palestinian women's awareness of and resistance to their own unjust, unhealthy and stagnant living conditions. Prior to mobilizing on a national scale, Palestinian women participated in organized activities primarily within the context of charitable organizations, as these functions fit well within the expected role of women in the Palestinian Arab culture. Women began organizing formally as early as March 1978, almost ten years before the *intifada*, when the first grassroots women's committee was established under Israeli occupation. However, for political and ideological reasons, this popular women's movement – the Palestinian Union of Women's Work Committee (PUWWC) – did not stay unified for long, and split into four groups along party lines.

It was the Palestinian *intifada* that provided dramatic momentum to these women's committees, for from the very beginning women participated openly in resistance activities against the Occupation under their umbrella. When curfews were common and schools were closed, neighborhood and popular women's committees concentrated on providing social and health services and community-based education. At the end of 1988 these committees were declared illegal by the Israeli authorities. The committees then began to address themselves to larger issues, uniting under the Higher Women's Committee (HWC) and planning future state structures as well as the role of women within these structures.

This cross section of politicized Palestinian women now forms a core of experienced leaders with broad-based support throughout the cities, villages and camps within the Occupied Territories, who are now ready to focus on the environment as a rallying point for women working to establish a Palestinian economy based on sustainable development. Most important for the future is to ensure that Palestinian women's activities – not only with regard to the environment – avoid factionalism and welcome all women, including those with no political affiliation, to work together towards the establishment of a democratic state.

Since it is women who are most affected by environmental problems, Palestinian women are likely to emphasize the protection of nature through the development of parks and tourist sites and through the protection and development of good, clean water resources. As consumers, Palestinian women can demand control over pesticide use, so that there will be fewer chemicals in foods, and encourage organic farming in the Occupied Territories. Palestinian women are expected to take an especially strong stand on the issues of sewage and solid waste management, as the current situation in the Occupied Territories has become unbearable.

All Palestinians – men and women – who are and will be working toward the establishment of an independent and free Palestinian entity need to remember that promoting development and protecting the environment are complementary. At the same time, any efforts at improving the general standard of living and the environmental awareness of the population as a whole in the Occupied Territories must at all times take into consideration the wide disparities among the problems affecting the women in the Occupied Territories. It must be remembered that while the poor are least able to escape environmental problems, the poor also benefit the most from environmental improvements.

Women's awareness has a special positive effect on the environment. Research carried out by the World Bank suggests that investment in education – especially for girls – is the single most important way of breaking the cycle of poverty, excessive population growth, and environmental degradation. According to the World Bank, because education does so much to ease poverty and slow population growth, by increasing women's awareness, it is one of the best environmental policies a country can have (World Bank 1992).

In the Occupied Territories, women's awareness is not so much an official policy as it is a necessity and a fact of life forced by the difficult conditions of the occupation. While this awareness has been achieved through years of difficult conditions and struggle, it also gives the Palestinians particular impetus and particular potential to build a society that will reinstate the harmony between human beings and their environment which existed for centuries before the intrusion of occupiers.

BIBLIOGRAPHY

Abu El Haj, S. (1992) 'Insecticides and health in the West Bank', in *Proceedings of Workshop on Environment, Land and Water Establishment*, Jerusalem.

Abu Safieh, Y. (1992) 'A brief description of the environmental aspects in the Gaza Strip', in *Proceedings of Workshop on Environment, Land and Water Establishment*, Jerusalem.

Al-Khatib, N. and Assaf, K. (1993) 'Palestinian water supplies and demands', in *Proceedings of The First Israeli–Palestinian International Academic Conference on Water*, Zurich, Switzerland: Elsevier Science Publishers.

Al Quds Daily Newspaper (January 20, 1993) 'Israeli settlers are responsible for the destruction of the environment and natural scenery in the West Bank', Jerusalem (translated from Hebrew).

Assaf, D. (1989) 'Palestinian housing and living conditions in the West Bank: A descriptive analysis', Masters Thesis in Urban Planning, University of Washington, Seattle, WA.

Assaf, S. and Assaf, K. (1985) 'Food security in the West Bank and Gaza Strip', for ESCWA/FAO Joint UN Division and AOAD of the Arab League, UN Publication, October.

—— (1986) 'The water situation in the West Bank and Gaza Strip', in *Water Resources and Utilization in the Arab World*, Kuwait City, Kuwait: The Arab Fund, 93–136.

Barghouthi, M. and Daibes, I. (1993) 'Infrastructure and health services in the West Bank: guidelines for health care planning', Ramallah, West Bank: Health Development Information Project.

Fakhouri, M. (1993) 'The various sources of pollution in Deheisha Refugee Camp', *Al-Asbou' Jadeed Magazine*, May 2 (in Arabic).

Haddad, M. (1992) 'Environmental conditions in the occupied West Bank', in *Proceedings of Workshop on Environment, Land and Water Establishment*, Jerusalem.

Heiberg, M. and Ovensen, G. (1993) 'Palestinian society in Gaza, West Bank and Arab Jerusalem: a survey of living conditions', FAFO Report 151, Faleh Hurtigtrykk: Oslo, Norway.

Kittaneh, O. and Hassan, A. (1993) 'The Palestinian electricity sector: guidelines for reconstruction and development', Center for Engineering and Planning, Ramallah, West Bank.

Land Research Committee (1993) 'The agricultural violations through the five years of the *intifada*', Jerusalem: Arab Studies Society, February.

Lipman, B. (1988) *Israel: The Embattled Land: Jewish and Palestinian Women Talk about their Lives*, London: Pandora.

Palestine Human Rights Information Center (1993) *Human Rights Update*, May 20, Jerusalem: Arab Studies Society.

Safi, J. and El-Nahhal, Y. (1992) 'Mutagenic and carcinogenic pesticides used in the Gaza Strip agricultural environment', in *Proceedings of Workshop on Environment, Land and Water Establishment*, Jerusalem.

Shqair, A. (1992) 'Forests and natural parks in the West Bank and their role in promoting internal tourism', in *Developmental Issues – Tourism in Palestine*, 2, 2: 18–23 (in Arabic).

—— (1993) 'Summary of the environmental situation in the occupied Palestinian territories', Paper presented in the first regional conference on the regional conditions for successful environmental policies in the Arab World, Jordan: Yarmouk University, April 3–5.

(UNRWA) United Nations Relief and Works Agency (1992) *Guide to UNRWA*, UNRWA, Vienna, Austria: Public Information Office.

World Bank (1992) 'The World Bank and the environment fiscal 1922', Washington, DC: The International Bank for Reconstruction and Development.

11

A FEMINIST POLITICS OF HEALTH CARE

The case of Palestinian women under Israeli occupation, 1979–1982

Elise G. Young

The concept of health is historically specific, gendered and, in twentieth-century Palestine, connected to the processes of colonization and the formation of the modern nation-state. This chapter addresses the effects of Israeli military occupation in the West Bank on the health of Palestinian women and on the way in which health is defined by some Palestinian health professionals. I argue that self-determination is a critical condition for women's health, and that self-determination depends upon redefining both the nationalist struggle and health and health care systems from a feminist perspective.

Definitions of health and the development of health care systems in present day Palestine emerge in the context of the struggle for control of nature as resource and of women as resource that has characterized West/East relations of imperialism and colonialism. In this case, violent means of controlling nature and of controlling women by the Israeli military government result in unhealthy conditions and modern diseases, and in denial of health care services to the largest segment of the Palestinian population living in poverty – women and children. In addition to its direct effects on the health of Palestinian women, the Israeli military occupation both exploits and affects indigenous systems of control of women. In turn, Palestinian women's resistance to Israeli military occupation has stimulated debate about the meaning of self-determination for women within Palestinian society. One important outcome of this debate has been the formation of the Women's Health Project of the Union of Palestinian Medical Relief Committees, discussed in this chapter.

In the history of West/East relations, control of nature, of women, and of colonized peoples – that is, of 'despised' populations – has been supported by anti-feminist, racist ideologies such as social Darwinism, which create and link concepts of gender and race justifying imperial rule. While they hoped to solve the historical suffering of Jews as a despised population, to achieve their goals Zionist leaders engaging in diplomacy with the then 'great powers' also utilized and were supported by these historical precedents.[1] For example, the use of Palestinians as a cheap labor force for the capitalization of Israeli society is linked to processes of both war and industrialization which disrupt sensitive environmental systems. Under occupation, and as the 'despised race,' all Palestinians are treated like 'women': inferior and undeserving of rights: untrustworthy: irrational. This chapter, therefore, treats the issue of health for all Palestinians living under occupation as a feminist issue. More specifically, however, this chapter focuses on the health status of women for two reasons. At the bottom of the gender–race–class hierarchy, the health status of women is particularly at risk. But the health status of women is also especially endangered because policies that endanger women's health – for example, control of women's reproduction – are also utilized within the indigenous Palestinian society, albeit differently, to control women. Direct and indirect alliances of men, which compete for control of women across ethnic–racial, and sometimes even class boundaries in securing internationalist patriarchy, are a critical health hazard to Palestinian women living under occupation.

Thus the Union of Palestinian Medical Relief Committees and the Women's Health Project attribute the critical conditions affecting Palestinian women's health both to the Israeli military occupation and to the subjugation of women within Palestinian society. For example, malnutrition as a result of both 'occupations' is a major determinant of the health status of Palestinian girls: on one hand, control of water sources by the Israeli government contributes to malnutrition; and, on the other, girls suffer higher rates of malnutrition because mothers who lack state services, and thus are dependent in their old age on the economic support of sons, may prioritize boys' health.

Key determinants of women's health between 1979 and 1982 were: the establishment of the Union of Palestinian Medical Relief Committees (UPMRC) in 1979; the popular uprising of Palestinians against Israeli military occupation, *intifada*, which emerged in December, 1987; the Israeli military response to the *intifada*; and the UPMRC creation of the Women's Health Project, in 1988. The history of the effects of West/East imperial relations on the development and institutionalization of medical theory and practice in some areas of the Middle East will further illuminate current struggles within Israel–Palestine for control of health care.

THE FOUNDING OF THE UPMRC AND THE WOMEN'S HEALTH PROJECT

In 1979 Dr Salwa Najjab and her colleagues founded the Union of Palestinian Medical Relief Committees, in response to three critical constraints on the health of Palestinians living under occupation. First, the Israeli military places severe limits on Palestinian access to medical resources. Second, because the Israeli health care system emphasizes hospital services rather than primary health care, and hence curative rather than preventive services, even when these services are available they do not meet the needs of Palestinians living in rural areas. Third, Palestinians were in need of a health care system which would address the particular needs of women, who, with children, constitute 70 per cent of the population and are, according to studies by the Women's Health Project, more at risk of disease and dying and at a disadvantage, socially as well as medically, within Palestinian society.

Trained in the Soviet Union as a gynecologist, Dr Najjab had worked for several years in Maqassed hospital in East Jerusalem, where she started a clinic for women. She observed that Palestinians who traveled to Maqassed from rural villages were not benefiting from the health care system's focus on doctor–patient care within a hospital setting. Many could not afford prescribed medicines. Women often did not have the means to carry out treatment of sick children after returning home. Children died who could not reach the hospital in an emergency. In response, Dr. Najjab and colleagues organized the UPMRC, and began traveling to villages in mobile health clinics.[2] Since 1979 the UPMRC has created thirty health care centers in the West Bank and Gaza, mainly in villages where there are no other health services. The mobile clinics were the first stage in developing a primary and preventive health care system that provided affordable and accessible care on an ongoing basis. As the central providers of health care to their families, women attending the clinics especially benefited from community-based health care.

Dr. Najjab and her colleagues were concerned about findings pointing to the need to develop a community-based primary health care system that addressed health not only in relation to biology and to the material and stress-related conditions of occupation, but also in relation to gender and class status. Infant mortality rates for Palestinians far exceed those of Israeli Jews. In three Palestinian villages in the Ramallah District, 1981 figures cite 91/1,000 infant mortality rates, in dramatic contrast to Israeli Central Bureau statistics claiming 14/1,000 for Israel in the same period (UPMRC 1987: 8–10). In addition, malnutrition is a major health problem for Palestinians, as opposed to Israeli Jews. Furthermore, UPMRC and Birzeit University studies found higher infant mortality rates among Palestinian girls than among boys, as well as the fact that, as noted earlier, girls suffer a higher rate of malnutrition. Union of Palestinian Medical Relief Committees studies found that in the village of Biddu, for example, infant mortality for girls was 58/1,000, while it was 41/1,000 for boys. Palestinians living in poverty, particularly as refugees in camps, suffer diseases of poverty caused by contaminated water and poor sanitation: the major killer diseases for Palestinian children are gastroenteritis, summer diarrhea, respiratory diseases, and malnutrition (UPMRC 1987: 10–11). Studies by the UPMRC also found that women tend to suffer more disease than men, and that poverty exacerbated and extended by de-development of Palestinian society and policies of land confiscation, house demolition, and imprisonment predisposes women to disease. The major health problems of women include gynecological infections, arthritic pain, high incidence of diabetes, and a rising rate of cancer of all kinds, especially reproductive (Najjab 1989: 3–4). In addition, the level of anemia among pregnant and lactating women in the Occupied Territories is 50 per cent: 20 per cent of all mothers suffer from anemia (UPMRC 1987: 10). Furthermore, women attending clinics complain about domestic tensions affecting their health; about for example, anguish about a husband who, taking a second wife, treats the first wife badly, or about brothers attempting to control their sisters' social lives.

In order to reach women and to provide primary and preventive health care services to address women's specific health problems, in 1988 Dr Najjab initiated the Women's Health Project. The groundwork for the Women's Health Project was laid by grassroots women's committees which conducted studies focusing on rural women and their needs; in some cases members of these committees eventually became health workers for the Project. Health care is provided by women doctors and by women from villages in the locality of the clinics trained as health care workers through the UPMRC. Among the services provided are pre-natal and postnatal care; family planning; gynecology; cancer screening; and health education. Treatment by women doctors is a defining aspect of this health care system: women attending the clinics express appreciation for relationships with women doctors who are 'not only our doctors, but our friends and our sisters.'[3] As doctors, sisters, and friends, Dr Najjab and her colleagues observe first hand the health effects of Israeli occupation on women, and the effects of indigenous practices of control of women.

RESTRICTIONS FROM WITHOUT: HEALTH CARE AS AN INSTRUMENT OF WAR

From 1979 to 1992 the Israeli military has continued policies begun in 1967 of utilizing health as an instrument of war, despite United Nations documents denouncing such policies. Palestinian women suffer particularly as a consequence of these policies, because of constraints within indigenous society and within a sexist–racist global economic structure. For example, those least capable of reaching health services, because of inadequate income or inability to gain access to family income, are women and children. Inequities in distribution of health care by the Israeli administrative apparatus institutionalize sexist, racist and classist assumptions which equate 'Palestinian' and 'woman' and reduce both to manifestations of 'nature' needing to be tamed and controlled.

Since June 1967, the Occupied Territories have not been represented in state health policies, including provision of health insurance, coordination of various health sectors, containment and control of infectious and communicable diseases, control of the environment, inspection of public places, and long-term planning for the solution of health problems (UPMRC 1987: 18). According to a study by the West Bank Data Base Project in 1986, the average Israeli military government expenditure on health services in the West Bank and Gaza did not exceed 30 US dollars per person per year, in comparison with 350 dollars spent in Israel per Israeli. In 1985 the budget for one Israeli hospital was six times the amount assigned to all nine government-run hospitals in the West Bank (UPMRC 1987: 16–17). Villagers have only limited access to hospitals run by the Israeli government, either in the West Bank or in Israel. Although the population in the West Bank increased between 1974 and 1985 by 21 per cent, the number of hospital beds available fell 6 per cent. And 248 out of a total of 489 localities in the West Bank, including villages, refugee camps and towns, are without any form of modern health services (UPMRC 1987: 12–15, 19).

The Israeli civil administration's refusal to develop health systems in the Occupied Territories makes Palestinians dependent upon medical services in Israel; yet, at the same time, the Israeli government systematically excludes Palestinians from such services. At the beginning of the *intifada*, then Defense Minister Yitzhak Rabin reduced the number of hospital beds in Israeli hospitals available to Palestinians – utilizing health care as an instrument of war, to ensure further loss of life rather than to save lives. And since 1977 the Israeli government has progressively raised monthly fees Palestinians must pay for use of government health services; at the same time, because services have also been cut, many of those Palestinians who can afford to pay have stopped doing so. In the village of Biddu, for example, only 22 per cent of the people are insured and only 4 per cent regularly utilize any health services provided by the Israeli government. Of the total population of Palestinians in the West Bank and Gaza 70 per cent lack any form of health insurance and cannot pay for treatment because of high hospital costs (UPMRC 1987: 20–21). Finally, in spite of the dire need for physicians, there is no structural apparatus to employ Palestinian health professionals, especially given that the Israeli government hospitals in the Occupied Territories employ as few physicians as possible. The physician/population ratio in 1986 in the West Bank and Gaza was 8/1,000, compared to 28/1,000 in Israel and 22/1,000 in Jordan. During the last three to five years about 100 physicians in the West Bank have remained unemployed – while UNRWA (United Nations Relief and Works Agency) doctors must examine 70–120 patients in a period of 4–6 hours; that is, one patient every three minutes (UPMRC 1987: 12–15).

These statistics follow from the historical process, begun under the British and continued under Israeli rule, of incorporating Palestine through violent subjugation into a world economic system. Military occupation is a major health hazard affecting Palestinians' resources, choices, and chances for survival, and facilitating the use of health care as an instrument of war. By subsuming medical needs under political considerations, the Israeli government is consigning Palestinians to the lower classes in a developed capitalist system.

IMPRISONMENT AND DEPORTATION: THE 'TAMING OF THE SHREW'

Women are specific targets of policies meant to demoralize, dispossess, and disempower Palestinians in the Occupied Territories. These policies include humiliation of women arrested, through physical and sexual terrorism; deportation; and attempts to control reproduction.

One of the dramatic health implications of military rule for Palestinians is daily confrontation with armed soldiers. Thousands of Palestinians are currently detained as political prisoners, under health threatening conditions. From December 9, 1987 through March 31, 1992, Israeli soldiers killed 1,032 Palestinians and injured an estimated 121,096 Palestinians (PHRIC 1990, 1992). Children are the population most affected, and many suffer lifelong disabilities from gunshot wounds and/or beatings. Moreover, between December 9, 1987 and September 30, 1990, ninety-six women were killed by Israelis, some in demonstrations, some walking to school, some standing outside hanging out their washing.

The Israeli military have also arrested and detained more than 2,000 Palestinian women (PHRIC 1990). The experiences of Palestinian women in prison clarify the connection between militarism and sexual conquest, and the consequent threat to the physical and psychological health of women. While men are also tortured in prison, humiliation of women is an additional tactic meant to disempower Palestinian men, for whom masculinity is connected to control, most often discussed as 'protection,' of women. The link between race and sex is implicit in this statement by an Israeli soldier:

> We were standing at a checkpoint. Along came a good-looking woman, well dressed and proud, in a way that succeeded in making my commander nervous. Although he usually exhibited extraordinary self control and sensitivity, he suddenly became very angry and got into the 'taming of the shrew' syndrome. Her pride was interpreted as an assault against us. The fact that they are standing on their legs without being afraid seems like a revolt against us. It is necessary to humiliate them, not because we are sadists, but to make it clear who is the adult and who the child. It is not good that there is such confusion and the pyramid must be re-erected on its foundation. Such a woman, walking proudly, tears down the whole system.
>
> (PHRIC 1990: 102)

One tactic used to humiliate women is the handcuffing of pregnant women prisoners when they go into labor, a graphic example of 'taming' the 'animal nature' of 'the Arab' and a graphic expression of occupation as control of women's fertility. In prison, women also suffer ill health from denial of medical treatment and from harsh living conditions, including cold, poor sanitation, and poor nutrition, and they suffer detrimental health effects of sexual violence experienced in the course of interrogations.

In 1987, the Women's Organization for Political Prisoners (WOFPP), based in Tel Aviv, initiated a campaign to publicize tactics utilized against Palestinian women in Israeli prisons and detention centers. They report repeated instances of sexual retaliation, including threatening beatings of women's breasts and genitals. At the Russian Compound in Jerusalem, for example, a male official threatened one detainee with rape and asked her how many men she had slept with. 'Then he told me, "Stand in the corner and I will take your clothes off and do many horrible things to you . . . I will take your breast and put it on the edge of the table and punch it with my hands. I will make sure that you will not be able to be a woman any more."' Because she refused to talk she was forced to sit in a chair with something sharp in her back and her hands tied behind her. Water dripped on her head. She was not allowed to go to the bathroom for two days, or to have food or water. She wrote: 'I felt I was going to die, and that this was the end of my life' (Najjab 1989–1990: 2). In addition, often women are wounded when they are arrested, or they may have long-standing medical problems that require daily treatment, and medical attention for them is delayed, denied, or inadequate. Interrogation and torture leave women with severe physical and emotional wounds. Lack of nutrition, exposure to cold, and separation from family and friends wear women down; and while many resist with hunger strikes and with solidarity, medical attention remains, nonetheless, necessary.

The March 1992 Newsletter from the (Israeli) Women's Organization for Political Prisoners (WOFPP) further details the range of health problems experienced by women prisoners. An eighteen-year-old woman sentenced to life plus twelve additional years is kept in solitary confinement, allowed only one hour of exercise daily. A twenty-nine-year-old woman from the Bethlehem region suffers from nosebleeds and ear problems as she waits for an operation that has been postponed with no new date set. Another woman, suffering from kidney stones and losing her eyesight, is brought to a hospital: the need for an operation is confirmed, but no date is set. A twenty-two-year-old whose arm has been paralyzed since she was shot during her arrest has not had an operation. A twenty-two-year-old from Tulkarem, arrested on January 22, 1992, was interrogated in a detention center for fifteen days, including two three-day and one four-day non-stop interrogations. During this time she was kept in solitary confinement and verbally abused. She was taken to the cell of male criminals and told that unless she confessed she would be left there and raped by all of the prisoners (WOFPP 1992: 2–3).

The Israeli military also acts as a mechanism to control population growth, through the use of weaponry harmful to pregnant women and through policies of deportation. Miscarriages and intrauterine fetal deaths resulting from tear gas used in enclosed spaces, including in several instances in nurseries, are widely reported. In September of 1991 alone, thirteen Palestinian women miscarried after being tear-gassed (PHRIC 1990). In addition, in a pattern that has been escalating since the *intifada*, the Israeli government has been denying Israeli identity cards necessary for residency in the Occupied Territories and deporting Palestinian women. In June, July and August of 1989, for example, fifty-seven Palestinian women were deported and over two hundred children expelled with them or separated from their mothers (Keizer 1989: 9). Soldiers often round up women in the middle of the night, giving them no time to collect their belongings or notify relatives. These policies of separating families have continued despite a 1990 Israeli High Court recommendation against forced expulsion of family members.

Separating women from husbands and children creates stress with severe consequences for women's health. For, finally, women carry the major burden of responsibility for coping with the tragic effects on children of living through war. They are primary caretakers, providing nurturance, support, and physical help for children permanently injured. Even when hospitalized, Palestinian children are not safe from the Israeli military, for soldiers forcibly round up children from their hospital beds. During the spring of 1983, some 1,000 Palestinian schoolchildren were hospitalized for toxicity related to poisoning. Some of these cases may even have been caused by the psychological stress of war (Giacaman 1988: 121). Yet mothers, who may need to administer daily medications to injured children, often are prohibited access to medical facilities or may be separated from their children altogether. And health professional Rita Giacaman describes the case of one woman who, in response to her daughter's recurring nightmares, had to plead with an Israeli soldier to assure her daughter that he would not kill her or her mother (Giacaman 1988: 121).

NUTRITION, WATER AND WAR

Devastation to the environment from the continuing war for control of Palestine creates ongoing conditions which are dangerous to health, with particular consequences for women. Displaced from land and home, thousands of Palestinians have lost control over agriculture as a source of livelihood and of subsistence. Women have been injured or killed while returning to confiscated fields or while surreptitiously leaving homes when under curfew, attempting to find food and water for their families.

Moreover, under military rule and denied resources, women have major responsibility for caring for children under unsanitary conditions, with minimal protection from cold and women give birth in refugee camps and in villages where there is no running water.

Israel has also introduced highly technical agricultural methods with environmental consequences that affect health. Agriculture in the West Bank has grown, and involved widespread use of pesticides and overutilization of existing water sources and arable land (Whitman 1988). Groundwater quality has deteriorated because of the use of chemical fertilizers and pesticides, and because of seawater intrusion from overpumping domestic and industrial wastewater discharged without proper treatment into streams and wadis (channels through which water flows in rainy season) (Whitman 1988). Single crop production and intensive irrigation have caused soil erosion and increased salinity and pollution. Extensive well digging in the West Bank for Israeli use has lowered the water table, and many Palestinian wells and springs have dried up. An Israeli environmentalist predicts that by the year 2000 West Bank aquifers will be destroyed (Skutel 1986: 22–24). Palestinians in the West Bank and Gaza must obtain licenses to utilize wells and irrigation installations; their applications are frequently refused and their water quotas remain low. Palestinians have not been allowed to develop new wells for irrigation.[4]

Availability of water, and hence conservation of water sources, is critical to health as well as to the development of a viable economic infrastructure. Inequities following from Israeli control of water sources have serious consequences for the health of all Palestinians, and particularly for women. Water collection is a woman's chore, and it takes a woman on average two to six hours daily to carry water for a six-person household (Mansour and Giacaman 1984: 7). Rita Giacaman recounts the experience of Um-Ahmad (the mother of Ahmad), a patient in a clinic operated by the Charitable Society and Birzeit University who suffered stress-related health problems connected to the Israeli occupation and, especially, its water policies. Um-Ahmad's husband of twelve years was imprisoned by the Israeli army for a 'security offense,' with no formal charge or trial. Hardship in the face of inflated food prices – since she and her family did not own cultivable land – was exacerbated by the indigenous kinship structure, as her mother-in-law took charge of Um-Ahmad's life in ways which were often tyrannical. When the water cistern broke down, there was no money for repairs and Um-Ahmad had to carry water on her head, four to five hours a day from the village spring. Her ten-year-old daughter helped, and as a result fell behind in her classes. In the late summer months and autumn, the spring dried up and the family was forced to purchase water from a vendor in town. Meanwhile, the village elders applied for and were denied a permit to install piped water. They waited interminably to hear from the Israeli authorities. In the end they were not allowed control over their own spring water; they were told that 'the spring water, even the rain water, is the property of the State of Israel' (Giacaman 1988: 81).

Along with the physical stress of carrying water long distances, women like Um-Ahmad must care for children affected by unsanitary conditions and disease because of lack of water. A Birzeit study found a 41 per cent rate of malnutrition among children in the three West Bank villages studied, and identified the non-availability of piped water as a major cause. Children from households with no running water had the highest rate of malnutrition, while children whose households had an external source of running water were less susceptible to malnutrition (Giacaman 1988: 125). Furthermore, internal latrines affected nutritional status negatively, probably because of the health hazard in situations where adequate water is lacking (Giacaman 1988: 126). The type of malnutrition found among Palestinian children is particularly insidious, because the mild or moderate forms found 'can easily escape the attention of both parents and health care providers, and ultimately end up causing long-lasting or permanent damage because of the oblique nature of their presentation' (Giacaman 1988: 124). Finally children living in overcrowded conditions suffer a higher rate of parasitic infection, caused by a 'generalized contamination of the environment' rather than by specific household sanitary conditions (Giacaman 1988: 131). The rate of parasitic infection among schoolchildren in Gaza, according to a 1976 military study, was 50 per cent (UPMRC 1987: 11). The health of Palestinian women suffers, in turn, as they bear the brunt of attempts by the Israeli military occupation to control their bearing of and caring for Palestinian children.

RESTRICTIONS FROM WITHIN: MOTHERS OF THE NATION

Because one consequence of military occupation is a struggle for control of women's bodies, the Israeli military occupation heightens tensions within indigenous Palestinian society as women face additional pressure to fulfill prescribed roles. Palestinian women involved in nationalist movements have a range of perspectives about what nationalism means, and many have questioned traditional mores as utilized by the nationalist movement; yet, at the same time, many women continue to experience heightened restrictions, including physical violence, as men attempt to control them under the nationally charged conditions of occupation.

For example, the mental health of Nadia (pseudonym), a thirty-year-old Palestinian health worker trained by the UPMRC, was threatened by her brothers' attempts to keep her from going to work at a clinic. The Community Health Program, one of several community-based programs sponsored by Birzeit University, had initiated a study in the village of Biddu near Jerusalem which eventually transformed Nadia's life:

Some girls from Birzeit University came here to make a study so they went from house to house to do a questionnaire, so I helped them and that's how I came to know the people in that committee. Through this questionnaire I met with the Medical Relief people and when they came here with the mobile clinic I was told that they were going to open a permanent clinic in Biddu with the agreement of the health committee from the village who said they wanted them. They told them they wanted a village health worker in this center, a girl who the people respect.[5]

But Nadia's enthusiasm for her work at the center was punctuated by anguish. Recently she had gone to a picnic with the medical relief committee without asking her brothers. One brother who supported her working at the clinic had recently died and the others were adamantly opposed to her working outside the home. Because Nadia had gone to the picnic, her brothers were refusing to return to her father's house.

In Nadia's village few women are active in the *intifada*, and Nadia expressed envy of women in neighboring villages who had more freedom to participate. This is another reason why her work in the clinic is crucial to her. Her commitment to Palestinian self-determination is fulfilled through her work in health care. Nadia feels that her well-being depends upon resistance to restrictions both from within her home and from without; to patriarchy in two related, though seemingly opposed, forms.

While not consistent within Palestinian society, male control of women's bodies is for many Palestinian woman, as it is for women globally, the first obstacle to achieving self-determination. Palestinian researcher Rema Hammami documents a 'vicious campaign in Gaza to impose the *hijab* (headscarf) on all women. The campaign included the threat and use of violence and developed into a comprehensive social offensive' (Hammami 1990: 24–28). The Mujama al-Islami or Islamic Resistance Movement, Hamas,

> endowed the hijab with new meanings of piety and political affiliation. Women affiliated with the movement started to wear long, plain, tailored overcoats, known as shari'a dress, which have no real precedents in indigenous Palestinian dress . . . Here the hijab is fundamentally an instrument of oppression, a direct disciplining of women's bodies for political ends. The form itself is directly connected to a reactionary ideology about women's role in society and a movement that seeks to implement this ideology.
>
> (Hammami 1990: 25)

Hammami describes a situation in which a dress code was enforced, not chosen: women in Gaza organized and resisted, 'struggling on a daily basis to maintain their right to choose and their right to a better future' and asking for support from the United National Leadership (UNL), 'who waited one and a half years before addressing the hijab campaign' (Hammami 1990: 24). In this case, the reluctance of the UNL to respond may indicate the extent to which members consider the disciplining of women's bodies an intrinsic or at least a permissible aspect of nationalist struggle.

The disciplining of women's bodies is a health issue both because of the stress caused by threats and because of the actual physical harm sustained by women when men, as in this example, of the UNL, attempt to force women to wear the *hijab*. Self-determination clearly means for women wresting control of their bodies not only from the occupier from without, but also from the occupier within. In this sense, despite its diverse meanings, nationalist struggle contains a common thread threatening the health both of Palestinian women and, in forms varying in kind and degree, of Israeli Jewish women: the control of women's bodies.

While proscriptions concerning marriage and reproduction have particular meanings within specific historic and geographic contexts, for the most part male-dominated nationalist movements equate motherhood with patriotism, and focus on the imperative for women to become mothers of the citizens of the new nation. When loyalty to the state, or to a nationalist movement, becomes loyalty to a system of enforced motherhood, motherhood becomes the priority and the defining characteristic of 'woman'. Metaphors describing women's relationship to nationalism, and hence to motherhood, are rooted in the reduction of women to a nature which is 'selfless,' endlessly nurturing and supremely moral. The exemplary woman is the mother of martyrs, the backbone of the nation. She is an idealized 'mother' devoted to correcting the excesses and evils of men, whose warring natures lead them astray. The precondition of her loyalty is altruism. But, as Janice Raymond notes, 'Altruism has been one of the most effective blocks to women's self awareness and demands for self-determination, serving as an instrument to structure social organization and patterns of relationship in women's lives' (Raymond 1993: 51). In fact, the 'altruistic pedestal on which women are placed . . . is one way of glorifying women's inequality' (Raymond 1993: 58), and is a tool for limiting women's self-determination through control of their bodies.

Israeli military occupation heightens those elements within indigenous Palestinian society that support 'self sacrificing' motherhood as the defining characteristic of 'woman'. But their self-determination and, hence, their health depends upon Palestinian women's ability to determine 'self' through action in the world and through beliefs, values, politics. Realizing this, Dr Salwa Najjab and the UPMRC founded the Women's Health Project in order to provide health options specific to women's needs, including family planning. At this historic moment, for Palestinian women to decide whether or not and when to have children, and to decide how many children to have, necessarily defines nationalist struggle as the province of women as well as of men. Focus on family planning has important implications for the Union of Palestinian Medical Relief Committees as a structure supporting a Palestinian state. In other words, the state benefits not only from supporting women in having healthy children, but also from understanding that self-determination for women also means control over their own biology. The Women's Health Project becomes, then, a site for women to take action in the world, not in response to an imperative of service to men but in a way that supports self-determination for, and benefits, all women. From this perspective, the meaning of the 'nation' must shift and, along with it, definitions of health and organizations of health care systems must evolve. The tools available in this process reflect how difficult this action is.

HEALTH AND SEXUAL AND REPRODUCTIVE SELF-DETERMINATION

Women's bodies are occupied by enforced definitions of 'woman' that serve the needs, interests, and continued power of the male state. Those definitions may shift but, globally, androcentric social, political, and medical systems define 'woman' as either mother or whore. Within the context examined here, the Israeli military defines the Palestinian woman as 'whore', as the woman whose defiance threatens the system; while Palestinian nationalism defines her as the saintly mother. The imperative within Palestinian society for women to become mothers is reinforced by this dichotomy. Thus, along with tensions resulting from competition between groups of men and women for control over definitions of 'woman', two critical issues impact on the decision-making process for women looking for an assured form of contraception: one is the means by which women are conditioned and forced to choose motherhood; the second is the control of contraception and the effects of various methods of contraception on women's health.

For women who choose to bear children, as well as for those who do not, health depends upon the availability of resources that allow for both options without harmful consequences. Tactics of the Israeli military – like shooting tear gas at pregnant women – can subvert the choice of Palestinian women to become mothers. As discussed earlier, male-dominated Palestinian nationalist movements also attempt to control women's bodies through the way in which they define nationalist struggle. In both cases, control of women's bodies assumes, as Janice Raymond notes, men's right of sexual access to women:

> Reproductive access to women is made easy and expected because of men's right to sexual access. Reproductive abuse of women's bodies is accepted as normal, because sexual abuse has paved the way. Reproductive technologies are the next step in male ease of access to women and the abuse of women's bodies under the guise of scientific

advancement and therapy. Women are required to spread their legs too frequently for medical probing and penetration.

(Raymond 1993: xxv)

In broader terms, the Israeli Occupation constitutes a stage in the historical processes of colonial penetration of Palestine, including medical probing in the form of experimentation on women, as empires have sought control over resources, trade routes, and regions. Women in Palestine–Israel were among those experimented on in the early stages of development of the Intra Uterine Device (IUD) between 1930 and 1957.[6] Thus experimentation on women's bodies is an aspect of colonization, carried out through the domestication of women, through control of their reproduction. This is the historic and current context within which the Women's Health Project attempts to define and to take control of women's health.

Contraceptive methods available to women reflect the misogyny of medical practices which require women to 'donate' their bodies in the name of 'progress', as it has been defined and instituted by medical technology. The complexities of this situation for the Women's Health Project are underscored in the testimony of one village health care worker who received a phone call from a seventeen-year-old woman interested in family planning. This woman rejected the option of the IUD because she has been told by her neighbor that it would produce infertility. The health care worker responded by assuring her that she could remove the IUD when she wanted a baby, and by talking about the problems of husbands who refuse to use condoms and don't want their wives to have an IUD, without confronting this woman's concerns about the dangers of the IUD. Medical ethicist Janice G. Raymond notes: 'Contraceptives like the Dalkon Shield, the pill, and Depo-Provera have all been the subject of public debate in the United States. Some, like the Dalkon Shield, have been proven to the satisfaction of the "experts" to be more recognizably dangerous than the pill' (Raymond 1993: 117). The number of women world-wide using IUDs rose from 15 million in 1978–1979 to 70 million in 1986–1987, despite the fact that their use poses serious health problems to women by significantly increasing the risk of inflammatory disease, infertility, and ectopic pregnancy (Williams 1986: 191). In addition, because the IUD can be used by a woman without her husband's knowledge it makes it possible to sidestep the more difficult task of convincing men of women's right to self-determination. As long as women feel constrained in relation to the power of their husbands in determining their reproductive lives – a constraint which is reinforced under conditions of military occupation in Palestine – they are in danger of choosing contraceptive methods that may be dangerous to their health.

One consequence of modern medical systems' management of women's choices is that women's limited options cancel out the 'freedom' implied by choice (Raymond 1993: 45). In criticizing the presentation of new reproductive technologies as an individual woman's choice, rather than as violence against women, Janice Raymond asks, 'is choice the real issue, or is the issue what choices and in what context selective women's choices ... are fostered?' (Raymond 1993: x). Environmentalist H. Patricia Hynes points out that barrier methods like the cervical cap and diaphragm, globally downgraded by the medical establishment, need a resurgence: 'Unlike hormonal contraceptives (and the IUD), they cause no known side effects. They are the safest of all reversible contraceptives in terms of mortality risks to women. They do not cause any delay or risk in fertility after cessation of use' (Hynes 1991: 47–52). Widespread use of barrier methods would not, however, serve the interests of science or capitalists as they attempt to control the reproductive choices available to women. To the extent that the medical options of grassroots health care movements such as the Women's Health Project remain tied to pharmaceutical manufacturing companies and multinational corporations concerned with profits rather than with women's health, these options also remain tied to the forces that have brought about Israeli military occupation of Palestine.

RACE AND SEX:
THE WEST/EAST POLITICS OF MEDICAL PRACTICES

European colonialists have consistently utilized medical practices and practitioners to control indigenous populations and land bases. Medical practices and health care systems can reinforce or 'carry' racist–misogynist ideologies through the ways in which they are developed, instituted, and distributed. Obstruction of access to and control of health care of Palestinians under Israeli occupation is one example of this colonialist dynamic.

The historical relationship between Arabic and European medicine has also been shaped by colonialist history. It is a historical relationship that clarifies the connection between, on one hand, development of medical practices and, on the other, sexist–racist ideologies supporting colonialist penetration of the Middle East and, hence, supporting Israeli military occupation of Palestine.

Rooted in Judeo–Christian–Islamic traditions and Galenic (Greek) medicine, Arabic and European medicine, prior to the nineteenth century, defined health as 'proper balance' of the humoral makeup of the person at risk (Gallagher 1983: 13).[7] The role of the healer was to correct imbalance (Gallagher 1983: 8). Both Europeans and Muslims identified miasma (corruption of the air caused by putrefying matter or decomposing bodies), contagion, and astral influence as causes of epidemic disease. Thus European and Muslim doctors incorporated each other's practices (Gallagher 1983: 100). But, according to historian LaVerne Kuhnke, economic penetration of the Middle East by Europe in the nineteenth century brought about a new dichotomy between European medicine and Arabic medicine. European medicine became authoritative, legal, desirable; while Arabic medicine, a medicine of the 'common' people, lost status and became semi-legal and less desirable (Gallagher 1983: 100).

European medical theories, originally derived from Galenic–Islamic sources, were increasingly influenced during the nineteenth century by the use of the microscope, underscoring the new trend to view organisms as atomized 'parts'. LaVerne Kuhnke notes that 'an emphasis on specificity in disease causation, prevention, and cure became a hallmark of cosmopolitan medicine' (Kuhnke 1990: 2). In turn, British and French colonialists used the new definition of disease as 'pathenogenic effects on the body of specific microorganisms' to reinforce notions of the 'alien' and 'exotic' as 'invasive foreign elements' (Kuhnke 1990: 1). They utilized medical science to construct sex and race, subsuming female 'identity' under biological cycles and reproduction and characterizing Arabs, or 'Orientals,' as (like women) by 'nature' deficient in character, passive, backward, irrational, and incapable of taking control over their own health. European medical practitioners – who were also politicians heavily invested in commerce – claimed the right of access to women and, simultaneously, to land bases and their indigenous populations (Gallagher 1983).

Both Nancy Elizabeth Gallagher's study of medical practices in Tunisia between 1780 and 1900 and LaVerne Kuhnke's examination of nineteenth-century Egyptian public health policy show how the French and the British used medical practices and public health policies to consolidate economic control of Tunisia and Egypt.[8] In response to a series of devastating cholera epidemics in both countries, British and French officials focused on acquiring control of quarantine systems in order to control trade. Ships suspected of carrying the disease, for example, were held up in ports. Historians writing from the perspective of the colonizer generally insist that French consuls had to take over quarantine management in Tunisia because of the Bey's heedlessness in the face of epidemic diseases; but in fact the French acted on behalf of vested economic and political interests (Gallagher 1983: 40).

The Israeli military conquest of Palestine continues these historical processes: racist–misogynist belief systems conflating the terms 'woman', 'native' and 'nature' with 'passive,' 'irrational,' and 'inert matter' inform the struggle for control of medical practices, medical systems, and definitions of health in Palestine today. Israeli control of definitions of health and health care practices is rooted in the reduction of 'Palestinian' to a sexist–racist concept of 'the Arab.' Furthermore, as Gallagher and Kuhnke illustrate in the cases of Tunisia and Egypt, in Israel–Palestine it is socio-economic and political factors that most centrally determine the course of disease and the development of effective treatment. In Palestine in 1992, high infant and childhood mortality rates among Palestinians reflect problems of malnutrition, infection, and lack of sanitation as a result of Israeli policies of development in the Occupied Territories. Concern for the health of 'the Arab' is subsumed under economic and political mandates, rivalries, and systems. Before the *intifada*, for example, decisions about whether or not a Palestinian should or would be hospitalized in Israel or abroad were made by the civil administration and medical committees. However, because of heightened tensions since 1987, new administrative procedures have been instituted which transfer this power to the civil administration's treasury; to administrators with no medical qualifications, whose considerations are instead economic and political (AIPPHR 1990: 35).

Another parallel between the history of European control of medical practices in Egypt and Tunisia and Israeli control in Palestine is striking. LaVerne Kuhnke concludes that the matrix of single patient–doctor care and curative medicine developed in tandem with biomedical technology threatened society, displacing preventive health care and shifting the focus away from environmental control methods, such as the development of more refined sanitation systems, which had been critical to obviating the force of epidemic diseases (Kuhnke 1990: 163–164). Similarly, Rita Giacaman concludes about the Palestinians in occupied Palestine, as does Kuhnke in her study of medical practices in Egypt, that the biomedical view of health and medical care, with its focus on individual biological variables, does not adequately address indigenous health problems. As discussed earlier, Palestinians need an affordable and accessible primary and preventive health care system that will also respond to the health consequences of occupation. Ill health in occupied Palestine is a consequence of socio-economic political conditions resulting from inequalities of gender, race, and class, as instituted both from without and from within. Indigenous health practitioners who rely solely on the biomedical model in developing a health care system are in danger of reinforcing those inequalities. In Tunisia and in Egypt, health care systems during the period studied by LaVerne and Kuhnke, were ultimately shaped by those most economically and politically powerful, and least by those 'who need help the most – the invisible rural poor and women' (Kuhnke 1990: 156). The Women's Health Project of the Union of Palestinian Medical Relief Committees is seeking to reverse this historical process.

CONCLUSION

Control of women's bodies is central to the struggle between the Israeli military and Palestinian health practitioners for control of health. The Israeli military occupation of Palestine relies on colonialist constructions of sex and race to define 'health' and to set health care priorities as a means of controlling the targeted local population. Ill health is the inevitable consequence of the specific conceptualization of 'woman,' 'nature,' 'foreigner', which is propagated by colonialist thinking and which justifies occupation of Palestine. In an attempt to bring an end to occupation of the West Bank, the Union of Palestinian Medical Relief Committees recognizes that it must also bring an end to occupation of the Palestinian 'body.' The creation of the Women's Health Project signals recognition of the interconnections within Palestinian society between ill health and the subjugation of women, and between ill health and poverty, and of the processes of colonization employed by the Israeli military and its state.

Both sexualized humiliation of women in prison as a form of punishment and nationalist imperatives utilized to control women's bodies are examples of colonization, or 'seasoning', of women, as explored in this chapter. The effects of 'seasoning' women to support colonization processes (whether within the indigenous social system or by the Israeli military) constitute a critical, and for the most part unacknowledged, aspect of women's health.[9]

Feminist re-visioning of the struggle for self-determination brings into the foreground underlying constructs that must be transformed to ensure health – of women; of nature; and of men. 'Health' must be understood as reciprocity in social relations and between systems of social organization and nature, and as dependent upon the cultivation of tools for survival which preserve rather than destroy life dependent resources. The support of feminists internationally in uncovering the crimes of the medical establishment against women, and in devising strategies to ensure the conditions necessary for women's health, will thus be a key factor in the realization of the goals of the Women's Health Project and the Union of Palestinian Medical Relief Committees.

NOTES

1 For an in-depth discussion of this point see E. Young (1992) *Keepers of the History*.
2 This information is based on a series of interviews with Dr Salwa Najjab, conducted in the offices of the Union of Palestinian Medical Relief Committees in East Jerusalem, December/January, 1992.
3 'It's very important for us to have a woman doctor. We feel happy that we have a woman doctor, woman to woman. We can talk about our problems to a woman much more easily without any border between us. We feel that she's not only our doctor, but also our friend and our sister.' Interview by Dr Najjab with a woman attending the clinic in Biddu, January, 1992.
4 See U. Davis, 'Settlements and politics under Begin', *Middle East Report*, 1979, vol. 9 no. 5: 12–18. See also J. D. Dillman, 'Water rights in the occupied territories', *Journal of Palestine Studies*, 1989, vol. 19, no. 1: 46–71, and H. J. Skutel, 'Water in the Arab–Israeli Conflict', 1986: 22–24
5 Interview conducted by Elise G. Young and Dr Salwa Najjab at Women's Health Clinic, Biddu, December, 1991.
6 See N. Williams (ed.), *Contraceptive Technologies, 1986–1987*, 1986: 191.
7 Galenic medicine viewed disease as an imbalance of the four humors of the body: hot, cold, moist, and dry; blood, mucous, yellow and black bile were the primary elements in the balance.

8 See Gallagher, N. E. (1983) *Medicine and Power in Tunisia, 1780–1900*, and L. Kuhnke, (1990) *Lives at Risk*.
9 See K. Barry (1979) *Female Sexual Slavery*. Barry coins the term 'seasoning' to describe the processes utilized to ensure women's loyalty to men.

BIBLIOGRAPHY

AIPPHR (1990) The Association of Israeli–Palestinian Physicians for Human Rights, *Annual Report*, Tel Aviv.

Amit, Y., Lerman, J. and Marton, R. (1991, May–September) Activity Report, Tel-Aviv: The Association of Israeli–Palestinian Physicians for Human Rights.

Anees, M. A. (1983) *Health Sciences in Early Islam*, Vol. I, Sami K. Hamarneh (ed.), Noor Health Foundation and Zahra Publications.

Antonious, S. (1980) 'Prisoners for Palestine: a list of women political prisoners', *Journal of Palestine Studies*, 9, 3: 29–80.

Avramovitch, D., and Marton, R. (1990) *Association of Israeli–Palestinian Physicians for Human Rights Annual Report*, Tel Aviv.

Barry, K. (1979) *Female Sexual Slavery*, New York: New York University Press.

Beinin, J. and Lockman, Z. (eds) (1989) *Intifada: The Palestinian Uprising Against Israeli Occupation*, Boston: South End Press.

'Campaign to Support Women Political Prisoners' (1989, July–1990, Jan.) *Najda Newsletter*, 30, 1: 2.

De Groot, J. (1989) '"Sex" and "race": the construction of language and image in the nineteenth century', in S. Mendusand and J. Rendall (eds) *Sexuality and Subordination: Interdisciplinary Studies of Gender in the Nineteenth Century*, New York: Routledge, 93–122.

Gallagher, N. E. (1983) *Medicine and Power in Tunisia, 1780–1900*, Cambridge: Cambridge University Press.

Giacaman, R. (1988) *Life and Health in Three Palestinian Villages*, Atlantic Highlands: Ithaca Press.

Hammami, R. (1990) 'Women, the Hijab and the Intifada', *Middle East Report*, 164–165, 20, 24–28.

Hulme, P. (1990) 'The spontaneous hand of nature: savagery, colonialism, and the Enlightenment', in P. Hulme and L. Jordanova (eds) *The Enlightenment and Its Shadows*, New York: Routledge.

Human Rights – A Compilation of International Instruments (1988) New York: United Nations, and Geneva: Centre for Human Rights.

Hynes, H. P. (1991) 'The pocketbook and the pill: reflections on green consumerism and population control', *Issues in Reproductive and Genetic Engineering*, 4, 1: 47–52.

Illich, I.(1976) *Medical Nemesis*, New York: Pantheon Books.

Keizer, B. (1989) 'Women in Black', *The Other Israel*, 38: 9.

Kuhnke, L. (1990) *Lives at Risk: Public Health in Nineteenth Century Egypt*, Berkeley: University of California Press.

Mansour, L. and Giacaman, R. (1984) 'The West Bank women's movement', *Off Our Backs*, 16, 3: 7.

Merchant, C. (1980) *The Death of Nature: Women: Ecology and the Scientific Revolution*, San Francisco: Harper and Row.

Mies, M. (1986) *Patriarchy and Accumulation on a World Scale*, London and New Jersey: Zed Books.

Musallam, B. F. (1983) *Sex and Society in Islam: Birth Control before the 19th Century*, Cambridge: Cambridge University Press.

Najjab, S. (1989) 'Notes on women and health in occupied Palestine', unpublished paper.
Palestine Human Rights Information Center (1990) 'An Israeli soldier's description of soldier and settler behavior in Jazalon refugee camp' [from *Ha'aretz*, March 11, 1988], Chicago: 102.
Palestine Human Rights Information Center (1992) Database project on Palestine Human Rights, Human Rights Update.
Palestinian Federation of Women's Action Committees, West Bank and Gaza Strip, *Newsletter* (1989): 4.
Peteet, J. (1986) 'Women and the Palestinian movement: no going back?', *Middle East Report*, 138, 16: 21–24.
Raymond, J. G. (1989) *A Passion for Friends: Toward a Philosophy of Female Friendship*, Boston: Beacon Press.
—— (1993) *Women as Wombs: Reproductive Technologies and the Battle over Women's Freedom*, San Francisco: Harper.
Rockwell, S. (1985) 'Palestinian women workers in the Israeli-occupied Gaza Strip', *Journal of Palestine Studies*, 14, 2: 114–136.
Roy, S. (1987) 'The Gaza Strip: A case of economic de-development', *Journal of Palestine Studies*, 17, 1: 56–88.
Sayigh, R. (1983) 'Encounters with Palestinian women under occupation', in N. H. Aruri (ed.) *Occupation: Israel Over Palestine*, Belmont, MA: Association of Arab American University Graduates, Inc, 269–293.
Skutel, H. J. (1986) 'Water in the Arab–Israeli conflict', *International Perspectives*, July/August, 22–24.
Turshen, M. (ed.) (1991) *Women and Health in Africa*, New Jersey: Africa World Press.
The Union of Palestinian Medical Relief Committees West Bank and Gaza Strip (1987) *An Overview of Health Conditions and Services in the Israeli Occupied Territories*, Jerusalem.
The Union of Palestinian Medical Relief Committees (1990) 'Objectives, Organization, and Activism, 1979–1990', Jerusalem.
The Union of Palestinian Medical Relief Committees (1991) *Report on the Women's Health Programme*, Jerusalem.
The Union of Palestinian Medical Relief Committees (1991) *Study on Nutritional Status of a Selected Sample of Under Five Palestinian Children*, Jerusalem.
Waring, M. (1988) *If Women Counted: A New Feminist Economics*, San Francisco: Harper.
Whitman, J. (1988) *The Environment in Israel*, Publication of the Israeli Ministry of Interior.
Williams, N. B. (ed.) (1986) *Contraceptive Technology*, 1986–1987, New York: Irvington Publishers, Inc.
Women's Organization for Political Prisoners (1992) *Newsletter*, Tel Aviv.
Young, E. (1992) *Keepers of the History, Women and the Israeli–Palestinian Conflict*, New York: Teachers College Press.

INDEX

Abdo, N.: on absence of Palestinian state 49; on Palestinian patriarchy 110; on Palestinian women's movement 35, 42, 51, 109; on stereotypes of women 122; on wearing of veil 108; on women's cooperatives 50
Abdul Jawwad, I.: on women in agriculture 81; on women and *intifada* 42
Abdul-Nasser, Gamal 111
abortion, and Israeli 'demographic war' 129
Abu Amr, Z., on Islamic radicalism 45
Abu Ayyash, A., on Palestinian fears 37
Abu El-Haj, S., on pesticides 174
Abu Safieh, Y.: on Israeli control of water 172; on land use in Gaza Strip 169; on sewage disposal 170–171
abuse, sexual 191
academics, meeting between Israelis and Palestinians 92
Accad, E., on militarism and violence against women 122
Acker, J., on underestimation of women's employment 140
African National Congress (ANC) 53
Agricultural Relief Committees 43, 55n.
agriculture: and access to water 172, 173; Agricultural Relief Committees 43; changes in under British mandate in Palestine 108; declining employment in 141–142; and division of labor 154; effect of Occupation on 80–82, 109–110; employment in 85n., 142–145; and environmental problems 172–173, 187; loss of Palestinian control of 186–187; in Occupied Territories 38; Palestinian economy based on 11–12; seasonal workers 38; subsistence farming 155; sustainable 166; symbolic importance of 81; of West Bank 68; women in 85n., 147

air pollution 175
Akka (Acre; Akko) 21, 32n.; establishment of charitable organizations 108; exodus of Palestinians from 164; Organization for Akka women 16, 32n.
al-Aqsa mosque 162n.
Al-Haj, M., on women in higher education 113–114
al-Hamdani, L., on Israeli prison system 76
Al-Khalili, G., on Palestinian women's movement 35
Al-Khatib, N. and Assaf, K., on water rights 169–170
Algeria 8, 34, 52, 103; employment of women 138; women's responsibility for 'motherhood' 44–45
Aloni, Shulamit 28
Anderson, B., on nationalism 63
annexation, 'creeping' 37
anti-Semitism 2
Antonius, S., on Palestinian women 37
Arab Ladies Association of Jerusalem 73
Arab states: lack of support for Palestinians 64; Palestinian attitudes to 111, 112–113; *see also* Egypt; Jordan
Arab Women's Congress of Palestine (1929) 37
Arab Women's League of Jerusalem 73
Arab Women's Society of Jaffa 73
Arab-Israeli war (1967) 4–5; and Palestinian nationalism 64–65
Arabs in Israel *see* Israeli Palestinians
Arafat, Yasser 84n.
Ard (honor), and men's sense of powerlessness 110
army: as 'agent of socialization' 128–129; development of women's consciousness of 95–101; force used against *intifada* 124; influence on social organization 5–6; and national identity 97; role of women in 97, 102, 104n.; self-image of 123; sexual abuse

199

INDEX

of Palestinian women 116, 124–125; and social mobility 97–98; status of soldiers 131; women's 'reserve duties' 129; women's responsibility to train sons for 99–101, 130; *see also* militarism
arrests 71, 72
Ashdod, exodus of Palestinians from 164
Ashrawi, H., on youth role in *intifada* 71
Askalan, exodus of Palestinians from 164
Assaf, S. and Assaf, K., on water rights 169–170, 172
Awartani, H., on agricultural employment 85n.

Barghouthi, M. and Daibes, I., on water supply 170
'Basic Law' (1981) 37
bastat (peddling) 8, 12, 147–163; of agricultural products 154–156; and capitalism 155–161; distribution of tasks 154–155; and educational levels 153; effect of Occupation on 147–148; implications of 148; and marital status 150–151; marketing strategies 156; men's employment in 153, 161n.; nature of goods marketed 151–152; non-agricultural goods 156–158; and self-devaluation 153, 160–161; selling of relief goods 151; social acceptance of 150–151; stages of 158; use of earnings 154; working conditions 152
Bedouins, changing labor patterns 151–152
Begin, Menachem 98
Beit Sahour (village), and *intifada* 71
Beit-Hallahmi, B., revisionist history 131
Ben Gurion, David, on 'demographic war' 129
Ben-Porath, A., on discrimination against Palestinian Israelis 113
Beneria, L., on underestimation of women's employment 140
Benholdt-Thompson, V., on capitalism 160
Benvenisti, M., on economic development 68
Benziman, U. and Mansour, A., on Israeli military rule 4
Bethlehem 149
Biddu: health clinic in 189; health service use 183; infant mortality rates 181
Birzeit University 109, 187, 188–189
Bisan Center for Research and Development 48
borders between Israel and Occupied Territories: harassment on 69, 75–76; sealing of 93, 94
Boserup, E., on seclusion and employment of women 139
Brand, L.: on Palestinian Arabs' flight 65; on Palestinian economy 11; on PLO 65, 75; on women's charitable organizations 74, 75

Britain 64; role in 'colonization' of Palestine 3; withdrawal from Palestine (1948) 3
British Mandate 35, 171; role of Palestinian women under 108–109
Brook-Utne, B., on non-violence 88
Brownmiller, S., on sexism and militarism 127

Camp David Accords (1978) 41
capitalism 162–163n.; and change in Palestinian social change 160–161; growth of under British mandate in Palestine 108; women's employment in *bastat* (peddling) and 157–161; and women's health 193
charitable organizations 36–37, 38–39, 41, 55n., 73–74, 112–113, 175–176; and political activism 74–75; under British mandate in Palestine 108–109; and women's committees 75; and women's traditional roles 74–75
Chazan, N.: dialogue with Mariam Mar'i 16–32; on *Reshet* 92; on women's protest groups 125
children: confrontations with soldiers 184; effects of war on health 186; and health care 179, 181–182; involvement in *intifada* 43–44, 71, 72, 77, 85n.
citizenship: Israeli 4; Palestinian Israeli attitudes to 115
class: and development of national loyalties 10; effect of experience of occupation 5; feminism where gender advantaged over 50; and gender 50, 108
clothing: significance of traditional dress 81–82; urging of dress code 189–190
Cobban, H., on PLO 65, 70
Cohen, Geula, on women's 'reserve duties' 129
Cohn, C., on militarism and violence against women 122
colonialism: and environment 12–13; and health care 179, 193–195; internal 68; Israeli military occupation compared with 5; and Occupation 10–14; use of gender and race 179–180; and Zionism 14n.
colonization: Arab opposition to 3; British role in 3; effect on Palestinian economic and social structures 37–38; and experimentation on women's bodies 192; and health care 195–196; Israeli occupation as 1; Occupation as internal 24; as 'seasoning' of women 196, 197n.
Communist Party (Palestinian) 39
commuting, by Palestinian workers in Israel 69, 140–141
compromise in government 28
confiscations of land 37, 62, 147, 186
Connor, W., on nationalism 63
conscience, Left as Jewish

200

INDEX

conscience 24–25
consciousness: of army and militarism 95–101; effect of Gulf War on 101–102; feminist 36; of Jewish women affected by *intifada* 125–126; and literacy 67; and mass communications 67; *see also* ideology
contraception 191, 192–193; barrier methods 192, 193; side-effects of IUDs 192
Cooke, M., on militarism and violence against women 122
cooperatives, run by women 50
creativity, strengthened by pain 20–21
culture: and nationalism 63–64, 65; Palestinian 65; women's role as preservers of 110–111
curfews 62, 70, 77, 152, 162n., 176, 186; and funerals 85n.

Dai LaKibush 30
Dajani, S.: on break-up of Palestinian social structures 35; on women and *intifada* 42
Darweish, M. 65; on *intifada* 70; on Palestinian patriarchy 76
deforestation 81, 166, 172; under Occupation 13
Democratic Women's Movement 111
'demographic war' 129, 130
demolition of Palestinian homes 77, 85n., 123; resistance to 123–124
demonstrations: against Gulf War 92; by peace movement 90; by rural women 74; move away from 112; suppression of 124
deportations 66, 71–72, 93, 184; of Palestinian women 186
desertification 172–173
detentions 71, 72, 76, 84n.; political prisoners 184; of women 184–186
Deutsch, Y.: on nationalism 63; on women's peace movement 96
development: access to infrastructures 68; and environment 176; underdevelopment of Palestinian economy 11–12
DFLP (Democratic Front for the Liberation of Palestine) 34, 39
dialogue: between Naomi Chazan and Mariam Mar'i 16–32; difference between women's and men's 23; and self-awareness 20; women's groups set up for 30–32
Diaspora 20–21, 65, 72, 107, 131; and feelings of powerlessness 97
difference, Palestinian sense of 66, 69
discrimination: against Palestinian Israelis 113; awareness of 112; based on gender, race and class 113, 116; in employment 40
disempowerment: of Jewish women 17; of Palestinian men 77, 110, 184
displacement, numbers affected by 164
dispossession 64–65, 66
Dura al-Qara 149, 156

economic deprivation 67–70
economy: informal 12; *see also bastat*; underdevelopment of Palestinian economy 11–12
education: in cities 168; closure of schools and universities 56n., 71, 85n., 176; and employment in *bastat* (peddling) 153; Israeli Compulsory Education Law 111; literacy rates among women 149–150; Occupation and 67, 114; popular education committees 43; in refugee camps 167; role in breaking poverty cycle 177; of rural women 74; and social views of women 149–150; system in occupied territories 161n.; under British mandate in Palestine 109; women in higher education 113–114; women's entry into 38; women's organizations and 40, 112
Efrat Committee on abortion 129
Egypt: Camp David Accords (1978) with Israel 41; medical practices 194, 195
Egyptian rule of Gaza Strip 74, 162n.
elections (1992): Left victory in 27–28; Peace Block in 32n.
elitism: and feminism 50; of Israeli army 97–98
embroidery 81–82
employment: in agriculture 85n.; of Arab women in Occupied Territories 138–146; in *bastat see bastat* (peddling); of black South African women 53; campaigns for women's rights in 40–41; changes in under British mandate in Palestine 108; discrimination and inequality in 40; effect of Occupation on 62, 80–82, 145; gender differences in 142–145; and harassment 83; informal 147, 148; membership of trade unions 50; of Palestinian workers in Israel 18, 69–70, 77, 83, 84n., 113–114, 140–141; peddling *see bastat*; professional and semi-professional jobs 139, 142–145; proleterianization of Palestinians 37, 38, 67–70; rates of pay 114, 118n.; sweatshop conditions 82; in West Bank and Gaza Strip 8; of women 38, 50, 55n., 81–82, 83, 111–112, 113–114, 139, 142–145
empowerment of Palestinian women 2, 17–18; conflict with private sphere 7–8; rural and working women 10
enemy, awareness of oppression by 116
Enloe, C.: on army as 'agent of socialization' 128–129; on sexism and militarism 122, 127
environment: and colonialism 12–13; and development 176; Earth Summit (1992) 166; effect of Occupation on 164–177; effects of Occupation on women 167–168; lack of local Palestinian authority 171; and modernization 12–13; noise and air

201

INDEX

pollution 175; use of pesticides and herbicides 173–175; water rights *see* water
Espanioly, N., and Sachs, D., on women's peace movement 92
ethnicity: and Palestinian social agendas and national liberation 48–49; *see also* race
experimentation on women's bodies 192

family: authority of father eroded 77; changes in under British mandate in Palestine 108–109; effect of Occupation on 9–10, 76–78, 83; extended (*hamoula*) 65, 107–108; forced separation of 186; and market economy 8; Palestinian women's role in 106; politicization of 76
family planning 190–191, 192–193
Farris, A. *et al*, on Palestinian economy 11
Fatah 34, 40, 65, 84n.
Fawzia, F.: on sexual oppression 117; on women's activism 108
Federation of Palestinian Women's Action Committees 89–90
femininity: construction of and Zionism 132–133; *see also* gender; masculinity
feminism: and anxiety about national security 98–101; application of to Palestinian women 47–48; consciousness of coming with *intifada* 36; and critique of 'security' culture 6; critique of war 89; developed by cooperation in peace movement 117; and elitism 50; fundamentalist backlash and 46; and gendered politics in Occupied Territories 46–51; in Israel 51; and marginalization 98; new agenda for Israeli women 94–95; opposition to Occupation 6–7; Palestinian women moving from theory to strategy 54; and Personal Status (Family) laws 47–48; previous Palestinian opposition to 47; and recognition of public–private contradictions 116; relationship of Israeli and Palestinian 51; secondary to national struggle 36–37, 41, 44–45, 103; on sexism and militarism 127; in South Africa 53; theorizing about social agendas and national liberation 48–51
feminist peace treaty 92
Flapan, S., on myths of birth of Israel 131
folklore: and *intifada* 83; Palestinian 65
Freedman, M., on security and feminism 129
freedom of movement 29
fundamentalism 2, 6, 8, 45–46, 189–190

Galilee, women's groups in 32n.
Gallagher, N.E., on development of medicine 193, 194
Gaza Strip: agriculture 142; Arab settlement in 4; education in 161–162n.; effect of Israeli colonization on 37–38; Egyptian control of 74; employment in *see under* employment; Islamic movement in 45; Israeli occupation of 6; land use 168–169; population density 168–169; refugee camps 164–165; sewage disposal 170–171; women's employment in *bastat* (peddling) 148, 149, 152
gender: and differences in employment patterns 139; loyalties of cutting across political loyalties 8–10; and malnutrition 180, 181; and national security 132–133; and nationalism 62–63, 73–82, 83, 103; social construction of 122, 126–135
General Union of Palestinian Women (GUPW) 34, 39, 75
Gesher (Jewish–Arab group) 31, 32n.
Ghazawneh, Haniyyeh 85n.
Giacaman, R.: on access to water 187–188; on children's health 186; on health care system 195; on Palestinian women's movement 35
Giacaman, R. and Johnson, P., on women and *intifada* 42, 123–124
Giacaman, R. and Odeh, M., on women's activism 73–74, 108
Gilliam, A., on feminism and national liberation 51
Ginat, J., on Palestinian women's role in family 106
Gluck, S.B., on Palestinian feminism 47, 50
Golan Heights 111
Graham-Brown, S., on Palestinian fears 37
grassroots participation 30, 42, 70; *see also* under charitable organizations; *intifada*; peace movement; women's committees
Green Line 19, 30, 69, 76
Grossman, on Palestinian Arabs' flight 65
Gulf War 32n., 54; and awareness of women's marginalization 101–102; effect on Israeli men 134–135; effect on Israeli women 88–89, 92–93; and violence against women 134–135

Hadad, Y.: on women and nationalism 75; on women's charitable organizations 74
Haddad, M., on land use on West Bank 169
Haider, A., on discrimination against Palestinian Israelis 113
Haifa: establishment of charitable organizations 108; Palestinian Arabs leaving (1947–9) 3–4, 164; university 113–114; Women in Black demonstrations 90, 93
Hamas (The Islamic Resistance Movement) 45, 189; Israeli support for 45; urging women to return to traditional roles 45–46
Hammami, R.: on dress code 189–190; on Islamic radicalism 45
hamoula (extended family) 65; in Ottoman period 107–108

202

INDEX

harassment 62, 70–72, 82; of Palestinian workers in Israel 69, 75–76; sexual 79–80, 116, 124, 184; of women employed in *bastat* (peddling) 152
Hasin, Amal Muhammad, murder of 121, 134, 135
Hasseini, Faisal 46
Hazleton, L.: on 'demographic war' 129; on women's identity 129
health care: and access to water 187–188; by Palestinian women 43; of children 186; and colonialism 195–196; community-based 181–182; conditions of women in prison 185–186; costs of 183; developing resources 10; and disciplining of women's bodies 190–191; East–West politics of medical practices 193–195; expenditure on 183; experimentation on women's bodies 192; as feminist issue 10, 11, 180; as instrument of war 182–184; and *intifada* 180, 195; limitations of biomedical model 195; mobile clinics 181, 189; physician/patient ratio 183; prioritization of boys 180; and self-determination 179; and use of pesticides and herbicides 173–175; women's organizations for 47, 175
Hebrew language, reinforcing gender inequality 6
Hebron 149
Hechter, M.: on economic deprivation 67; on internal colonialism 68; on nationalism 63
Heiberg, M. and Ovensen, G.: on Israeli military rule 166; on population density 165
Helman, Sara and Rapaport, Tamar, on women and Occupation 96
Hicks Steihm, J., on traditional image of Israeli women 97, 98
hijab (headscarf) 189–190
Hijab, N.: on employment of Arab women 138; on religion and law 48
Hiltermann, J.: on employment rights 40; on women and nationalism 75; on women's charitable organizations 74
Holocaust: and anti-abortion campaigns 129, 130; effect on survivors 22–23; and gender construction 129; and national security 131; and sense of helplessness 133
Holt, M., on Palestinian feminism 50
home: demolition of by Occupation forces 77, 85n.; humiliation by soldiers within 76–77; overcrowding of in refugee camps 169; Palestinian women in Israel restricted to 110; searches of in Occupation 76–77; violence in 134–135
honor: and men's sense of powerlessness 110; and response to sexual harassment 116
Hosier, R., on informal economy 12
house-arrest 77
Hurwitz, D., on Israeli feminism 51
Hussein, Saddam 32n.
Husseini, Faisal 56n.

identity: conflicts for Israeli Palestinians 1, 114–115; effect of Gulf War on 134–135; and myths of birth of Israel 130–131, 132; role of army in 97; *sabra* as symbol of national 132–133; shaped by family 106
ideology: and imposition of dress codes on women 189–190; internalisation of 9; myths of birth of Israel 130–131, 132; and national security 9, 98–101, 131–133; role of 'mother of martyrs' 42, 49, 74, 78, 190; of sacrifice for homeland 130; use of gender and race in colonialism 179–180; *see also* consciousness
immigration: effect on pre-1882 Jewish/Arab balance in Palestine 2–3; effect on traditional Palestinian social structures 35; to Israel 2; to Palestine 2–3
In'ash el-Usra (Family Rejuvenation Society) 38, 55n.
inequality, perpetuated by national security needs 131–132
informal economy 12; *see also bastat*; employment interrogations 79–80, 85n.
intifada 2, 8, 13, 21, 33, 36; aims of 55–6n.; and attitudes to resistance 35; and consciousness of gender inequality 118; dependent on grassroots organization 70; effect on Palestinian Israelis 115–117, 123–124; and folklore 83; force used by Israelis against 124; and fundamentalist backlash against women 45–46; goal of economic self-sufficiency 43, 161–162n.; and health care 180, 183, 195; historical development of Israeli women's role 89–91; intensification of Palestinian nationalism in 62, 66, 82–83, 117–118; international attitudes to 82–83; Jewish women and 17, 95–96, 125–126; meaning of 42; and non-violent protest 71; origins of 70, 84n.; popular appeal of 42–43; and sexual harassment 116–117; shaking up of Palestinian social structures 42, 43, 44–35; and significance of motherhood 78; and strikes 162n.; Unified National Leadership of the Uprising (UNL) 56n., 70, 152; and violence against women 127–128; and women's employment in *bastat* (peddling) 147–148; and women's issues 41; and women's role in agriculture 81; women's role in 33–34, 41–46, 49, 54, 55n., 71, 77,

INDEX

88, 89–91, 92, 172; youth involvement in 71, 72, 77, 85n.
Iraq, employment of women 138
irrigation 173, 187–188
Islam, fundamentalism 2, 6, 8, 45–46, 189–190
Islam, allegiance of majority of Palestinians 83n.
Israel: armistice agreement with Jordan 4; employment of women 138; establishment of (1948) 3, 35; recognition by PLO 89
Israel Central Bureau of Statistics 140
Israeli Arabs *see* Israeli Palestinians
Israeli army *see* army
Israeli Palestianians difference from other Palestinians 18–20; experience of military 21
Israeli Palestinians 14n.; in Ottoman period 107–108
Israeli women: internal conflicts among 5–7; opposition to Occupation 88–105
Israelis: alienation of 26; gap between personal identity and politics 26
Italy: seminar of Italian, Israeli and Palestinian women 94–95, 104n.; women peace delegates 90

Jabaliya refugee camp 154, 157, 158; women's employment in *bastat* (peddling) 149, 152; women's literacy rates 150
Jad, I., on women and *intifada* 42
Jaffa: establishment of charitable organizations 108; Palestinian Arabs leaving (1947–1949) 3–4, 164
Jammal, L., on Palestinian women's movement 35, 36
Jayawardena, K., on women and nationalism 75
Jeffords, S., on militarism and violence against women 122
Jerusalem: establishment of charitable organizations 108; events at al Aqsa Mosque 152; first Palestinian women's conference (1929) 108–109; Israeli unification of 37; Palestinian Arabs leaving (1947–1949) 3–4, 164; peace campaigners in 89–90; peace march (1990) 89–90, 92; Women in Black 93; women's employment in *bastat* (peddling) 149; women's organizations 73
Jewish Agency 3
Jirys, S., on Israeli land expropriation 109, 110
Joffe, E., on Palestinian nationalism 64
Johnston, R., on colonialism 11
Jordan: Arab settlement in 4; armistice agreement with Israel 4; and Palestinian economy 11–12; Palestinian employment in 68; Personal Status (Family) laws 48; PLO activities in 66
Jordanian rule 57n., 74, 75, 84n.
Joseph, S., on absence of Palestinian state 49

Kazi, H., on women and nationalism 75
Keddie, N., on employment of Arab women 138
Keizer, B., on deportations 186
Ketsiot detention center 72
El-Khalil, Samiha 38
Khreisheh, A., on employment rights 40
kibush (occupation) 126–127
Kittaneh, O. and Hassan, A., on refugee camps 165
Knesset 30, 32n.
Kuhnke, LaVerne, on development of medicine 194, 195
Kuttab, E.: on Algeria 34; criticism of Palestinian women's movement in Occupied Territories 52; on women's cooperatives 50
Kuwait, employment of women 138

labor market: informal 12; peddling *see bastat* (peddling); women's participation in 12; *see also* employment
Labor Party (Israel) 1992 election victory 93–94; peace agreement with PLO 93
land, confiscation of 37, 62, 109, 147, 186
land ownership 37, 62, 147, 186
Land Research Committee 172
land use: in Gaza Strip 168–169; on West Bank 169
language, and nationalism 63–64
law: application of religious law in 'personal' affairs 47–48, 57n.; and military rule in Israel 110
leadership, developing from grassroots 30, 42, 70
League of Nations 3
Lebanon: Israeli war in (1993) 93, 94; PLO activities in 66
Left (Israeli): election victory 27–28; and military 25; and Palestinians 25–26
Legrain, J.F., on Islamic radicalism 45
Lewis-Epstein, N. and Semyonov, M., on employment of Arab women 138, 139
liberalization: *intifada* sparking hopes for 44; Occupation and 21
liberation 109–110, 164
Likud Party 6; government (1977) 41; Palestinian response to 41; policy on Occupation 67
literacy 67, 75, 85n.; campaigns 74
Lod, exodus of Palestinians from 164
loyalties between Israeli and Palestinian women 8–10
Lustik, I., on Israeli military rule 110

INDEX

malnutrition: gender differences in 180, 181; long-term effects 188; and water supply 188

Mansour, A. and Giacaman, R., on access to water 187

marginalization: of alienated Israelis 26; effect of Gulf War on consciousness of 101–102; and feminism 98; increased by emphasis on security 5–6; of Jewish women 2, 5, 17, 91, 98

Mar'i, M. and Mar'i, S.: on changing role of Palestinian women 112; on men's sense of powerlessness 110

Mar'i, Mariam: childbirth analogy 17–19; dialogue with Naomi Chazan 16–32

market economy: and family roles 8; *see also bastat*; capitalism

masculinity: linked to control of women 184; militarized construction of 130–133; *sabra* as symbol of 132–133

mediatory role of Palestinian women 2

Medical Relief Committees 43, 55n.

Meir, Golda 83n.

Meretz party 32n.

militarism 2; dependency on gender construction 128–129; development of women's consciousness of 95–101; and sexism 126–135, 134; and status of women 102, 103; and violence 102; *see also* army

military rule 10; and deforestation 13, 172; economic and environmental effects 166–167; effect on Palestinians in Israel 109–11; similarities with military occupation 16

miscarriages, caused by tear gas 124, 167, 186

Mizrahi (Jewish) women 23

modernization, and environment 12–13

Mogannam, M., on Palestinian women's movement 35

Mohanty, C., on feminism and national liberation 51

moledet (homeland) 130

Morgan, R., on militarism and violence against women 122

Morris, B.: on myths of birth of Israel 131; on Palestinianans and other Arabs 64

mortality rates, for infants 181–182

motherhood: in Algeria 44–45; attempts to control 184; effect of Occupation on 9–10, 83; equated with patriotism 190–191; fears for sons in Israeli army 99–101; influence of Occupation on 76, 78; Israeli state control of 127, 128–129; and miscarriages caused by tear gas 124, 167, 186; and political involvement 78; politicization of 49, 95; redefinition in South African liberation movement 53; role of 'mother of martyrs' 42, 49, 74, 78, 190; significance for Israeli women 98; and social liberation 8; 'triple burden' of 44–45; and 'whore' figure 191; *see also* reproduction

Mothers against Silence group 104n.

myths: of birth of Israel 130–131; of sacrifice for homeland 130

Nablus, establishment of charitable organizations 108

Najjab, Salwa: formation of Union of Palestinian Medical Relief Committees (UPMRC) 180–182; formation of Women's Health Project 182, 190–191; on women's health problems 182

Nath, K., on Palestinian women's role in family 106

National Guidance Committee (NGC) (Palestinian) 42

national security *see* security

nationalism 64–67; and Arab–Israel War (1967) 64–65; causes of 63–64; and culture 65; defined in opposition to 'other' 82; and economic deprivation 67–70; and gender 33–34, 62–63, 73–82, 83, 103; and *intifada* 82–83; Israeli prison system and 72; and motherhood 190–191; and Occupation 64, 66–67, 82, 112–113; Palestinian women and 73–82, 106, 188–191; Palestinians not recognised as national group in Israel 110; as response to external threat 63–64; and restrictions on Palestinian workers in Israel 69–70

Nazareth 118

Nevo, J., on sexual harassment 116

Nizan, Gabi, on militarism and violence 134

noise pollution 175

nutrition 181–182, 185, 186–188

Occupation: applied to women's bodies 126–135, 135–136n., 135; and colonialism 10–14; compared with colonialism 5; development of Israeli women's opposition to 95–96; enslaving effects of 23–24; environmental effects of 164–177; and harassment 70–72; Israel's handling of 66–67; marginalizing Jewish women 17; and nationalism 64, 66–67, 82, 112–113; and need for self-determination 112; and political consciousness of Israeli Jewish women 103; positive aspects of 20–21, 29; sanctioning separation of communities 20; as shock to Palestinians 111; similarities with military rule 16; and status of Palestinians 1–2; undesirable effects on occupied 62; and violence against Israeli women 70, 134–135

INDEX

Occupied Territories: economic underdevelopment of 11–12; employment of women in 140–146; growth of women's committees in 41; *see also* Gaza Strip; Israeli Palestinians; Occupation; West Bank
oppression: individual and collective 116; sexual 116–117
Orientalism 135n.
Ostrowitz, R., on sexism and militarism 121, 127–128
Ottoman rule, role of Palestinian women in 107–108

Palestine Human Rights Information Center 172
Palestine Women's Union 73
Palestinian National Liberation Movement (*Fatah*) 34, 40, 65, 84n.
Palestinian state: absence of 49; declaration of (1988) 92
Palestinian women, internal conflicts among 5, 7–8
Palestinian Women's Working Committees, declared illegal 124
Palestinians, definition of 4
Palestinians in Israel *see* Israeli Palestinians
partition plan (1947) 3–4
patriarchy: and insecurity of Palestinian men under Israeli rule 110; women as control mechanism of 111
Peace Block 32n., 94
peace movement 30; Israeli feminists and 51; Palestinian Israelis working with Jewish women 117; representation in Israeli government 88–89; solidarity among women in 115
Peace Now group 98, 104n.
Peace Quilt 91
peddlers *see bastat* (peddling)
People's Party (formerly Communist Party) (Palestinian) 34
Personal Status (Family) laws (Palestinian) 47–48
pesticides and herbicides 13, 173–175, 176, 187; lack of safety measures 174–175; side-effects of 173–174, 175
Peteet, J.: on women and nationalism 74, 75; on women's charitable organizations 73–74
PFLP (Popular Front for the Liberation of Palestine) 34, 40
PLO (Palestine Liberation Organization) 34–35, 39, 56n., 84n.; acceptance of two-state solution 35; factions within 34; formation 65, 75; as government in exile 66; mobilizing resistance 66; peace agreement with Israeli government 93; power moving to grassroots 42;

recognition by Israel 89; women's representation in 34
political activism, by Palestinian women 2, 7, 9, 17, 30–31, 33–34
political activism *see also intifada*; peace movement; PLO; women's committees
Polk, W., on Jewish Agency 3
polygamy 48
population density: in Gaza Strip 168–169; in refugee camps 165
population growth: and 'demographic war' 129, 130; in Occupied Territories 141
prison system: and nationalism 72, 76; women in 76, 79, 184–186, 196; Women for Women Political Prisoners 91
production 156–163
proleterianization: of black South Africans 53; of Palestinians 37, 38, 55n., 108
protest *see* resistance
Prothro, E. and Diab, L., on Palestinian women's role in family 106
punishment, collective 62, 70
PUWWC (Palestinian Union of Women's Work Committee) 176

Qalandiya refugee camp 156

Rabin, Yitzhak 32n.; and health care provision 183
race, and politics of health care 193–195
Ramallah 149, 152, 154
Ramleh, exodus of Palestinians from 164
Rasheem, Ali Hashem 65
Ratz (Citizens' Rights Party) 32n.
Raymond, J.G.: on contraception 192–193; on women and altruism 190
Reardon, B., on militarism and violence against women 122
refugee camps 36, 75, 84n., 85n.; difficulties faced by women 165; disease in 181–182; environmental effects on women 167; Jabaliya 149, 152, 154, 158; population density 164, 165, 168–169; Qalandiya 156; scale of 164–165
refugees, proportion of Palestinians as 36, 109
relief committees 158, 165, 167, 179, 180, 182, 183, 190–191, 192, 195, 196
relief programmes, reselling relief goods 151
religion: and liberation of Palestinian women 50; and nationalism 63–64
religious fundamentalism 2, 6, 8, 45–46, 189–190
reproduction: attempts to control 128–129, 184; and 'demographic war' 129; and women's self-determination 190–193; *see also* motherhood
Reshet (Israeli Women's Peace Net) 30, 31, 92, 94, 125

206

INDEX

resistance: boycotting Israeli goods 71; changing attitudes to 35, 55n.; emergence of civilian resistance 42–43; mobilized by PLO 66; and violence 71; *see also* intifada

Reuven, on self-image of Israeli army 123

rights, employment 40–41

riots (1929) 74

Rishmawi, M.: on women's activism 108, 109; on women's employment 81, 82

Rogel, Einav, murder of 121, 134, 135

Rosenwasser, P., on violence against women 124

Rubinstein, D., on education and Occupation 67

Ruddick, S., on militarism and violence against women 122

rural women: and *bastat* 149–163; education 74; environmental responsibilities of 168; in Ottoman period 108; and Palestinian nationalism 73, 74

sabra, as symbol of national identity 132–133

sacrifice, nationalist 130; role of 'mother of martyrs' 42, 49, 74, 78, 190

Sadan, Haim, and abortion rights 129

Safad, exodus of Palestinians from 164

Safi, J. and El-Nahhal, Y., on pesticides 173

Sarid, Yossi 25, 26, 32n.

Sayigh, R.: on women and PLO 75; on women's designs 81

schools: closing of 85n., 176; under British mandate in Palestine 109

Seagar, J., on militarism and environment 13

sealing of borders between Occupied Territories and Israel 93, 94

searches: body searches of women 79; of Palestinian homes 76–77, 79

seclusion of women: and *bastat* (peddling) 150–151; and employment 139

security: Arabs remaining within Israel seen as threat 4; effect on women 5–6, 16, 28–29; effects of prioritizing 128, 131–132; feminist critique of emphasis on 6; and Holocaust defensive mode 22, 28; Israeli Jewish women and 95–101; and national anxiety 98–101; Palestinian attitude to 28–29; and social construction of gender 132–133; women seen as less of threat 28–29; and women's identity 9, 129–130; and Zionism 130–133

Segev, T., on myths of birth of Israel 131

self-determination: and health 179; and women's control of body 189–190, 190–193

Semyonov, M., on employment of women 138

Semyonov, M. and Lewin-Epstein, N., on Arab employment in Israel 69, 139, 140, 142

Senker, C., on sexual harassment 116

sewage disposal 170–171, 176

sexism, and militarism 126–135, 134

sexuality: connection of militarism and violence 126–127; effect of Occupation on 78–80, 83; and oppression of women 116–117; taboos on 117; threatened in prison system 185–186; violations of by Occupation 9–10; and women's 'honor' 79–80; and women's self-determination 190–193

Shadid, M.: on agriculture 80; on Palestinian economy 11

Shadmi, E., on militarism and violence 95

shaheeds (martyrs) 78, 85n.

Shaloufeh Khazan, F., on subordination of Palestinian women 111

shame 116

Shani (Women against Occupation) 91, 125

Shapira, A., revisionist history 131

Sharoni, S.: on Gulf War and violence against women 135; on Israeli feminism 51; on stereotypes of women 122; on women's protest groups 125

Shelley, T., on Palestinian workers in Israel 69

Shohat, E., on prioritization of national security 132

Shqair, A., on deforestation 172

Siniora, R., on employment rights 40

skilled workers 139, 142–145

Skutel, H.J., on water supply 187

Smith, A.D., on nationalism 63

Smock, A. and Youssef, N., on Palestinian women's role in family 106

social Darwinism 179–180

soldiers *see* army; military rule

South Africa 57–8n.; motherhood in 44–45; Pass Laws 58n.; women's politicization and national liberation 52–54

Statistical Abstracts of Israel 140

Statistical Yearbook of Labor Statistics 138

Sternfield, on sexism, racism and militarism 122–123

stone-throwing 43, 184

strikes 162n.

Strum, P.: on Palestinian feminism 47, 50, 52; on violence against women 124; on women and *intifada* 42; on women's cooperatives 50

Swisrky, S., on prioritization of national security 132

Syria 111

Talhami, G., on women's employment 81

INDEX

Taraki, L.: on Islamic radicalism 45; on women and *intifada* 42
taxes 183
tear gas, causing miscarriages 124, 167, 186, 191
Tehiya party 129
Tel Aviv, Women in Black demonstrations 90, 93
Tel-Aviv, protest groups 104n.
Thornhill, T., on sexual harassment 79
Tiberias, exodus of Palestinians from 164
torture 11
Touma, E., on Palestinian women's role in family 106
trade unions, women's membership of 50
trade, within *bastat* (peddling) system 155–158
Tunis Symposium, on employment rights 40
Tunisia, medical practices 194, 195
two-state solution 3–4

uneven economic development 11–12
Union of Palestinian Women's Committees 39–40
Union of Women's Committees for Social Work 40
Union of Working Women's Committees 39, 89–90
unions 39–40, 50, 89–90
United Nations 1947 Partition Plan 3–4; Conference on Environment and Development (Earth Summit) (1992) 166; Relief and Works Agency (UNRWA) 158, 165, 167, 183
United States, Israeli Jews identifying with 92
universities: Birzeit 109, 187, 188–189; closing of in Occupied Territories 43, 56n., 71; women at under Occupation 113–114
UNL (Unified National Leadership) 70, 189–190
UNLU (Unified National Leadership of the Uprising) 56n., 152
UNRWA (United Nations Relief and Works Agency) 158, 165, 167, 183
unskilled workers 18, 69–70, 77, 83, 84n., 113–114, 140–141; *see also bastat*
UPMRC (Union of Palestinian Medical Relief Committees), Women's Health Project 179, 180, 182, 190–191, 192, 195, 196
urban women, charitable organizations 73–74, 75

veil, adopted by upper-class women 108
villagers and villages *see* rural women
violence: against women *see* violence against women; amongst Palestinian youth 6; Israeli Palestinians and 21; and militarism 102; military occupation and 70, 121–135; Palestinian resistance and 71; and political despair 93; and Zionist ideology 132–133
violence against women 90, 94, 123, 134–135, 152; by Occupation forces 54n., 184; directed against Jewish women 6; and fundamentalist backlash 46; in home 47, 102; and *intifada* 127–128; as peace campaigners 90; as protest groups 125–126; sexual 116–117
Voluntary Work Committees 55n.
volunteer work *see* charitable organizations

Wadi Rahal 149, 151, 155
Walker, C., on South African feminism 53, 57–8n.
war: Arab–Israel War (1967) 4–5, 64–65, 164; of Israeli independence (1948) 109–110, 164; in Lebanon (1993) 93, 94; *see also* Gulf War; *intifada*
Warnock, K.: on Palestinian feminism 50; on sexual harassment 79–80; on women and *intifada* 42, 75, 106
waste collection and disposal 171, 176
water: access to 81, 147, 165; in cities 168; collection of by women 187–188; contamination of by soldiers 170; and health 187–188; for irrigation 173, 187–188; Palestinian access to 13, 172, 173; pollution of 13, 170, 174; proportions used by Israelis and Palestinians 169–170; in refugee camps 167; sewage disposal 170–171, 176; women's responsibilities for collecting 170
Weiss, D., on informal economy 12
West Bank: agriculture 68, 80, 142; Arab settlement in 4; conference on women (1990) 46–47; education in 161–162n.; effect of Israeli colonization on 38; employment in *see under* employment; establishment of new Jewish communities 67–68; Israeli occupation of 6; Jordanian control of 74, 75; land use 169; population distribution 168; provision of health care 183; refugee camps 164–165; rule by Jordan 84n.; sewage disposal 171; women's employment in *bastat* (peddling) 148, 149; Women's Studies Centers 48
West Bank Data Base Project 183
Whitman, J., on pesticides and herbicides 187
Women in Black 30, 90, 115, 125–126; group work on violence, militarism and feminism 90, 94–95; joint work with men and Palestinian women 94; as model for protest groups 90; protest vigils 96
'Women Go For Peace' event 92
Women and Peace Coalition 89–90, 102, 115,

208

INDEX

125; formation of 91; new feminist agenda 94–95; protest against Gulf War 92
Women for Women Political Prisoners 91, 115
women's centers 57n.
Women's Committee for Social Work 89–90
women's committees 10, 55n., 82, 114, 175; beginnings of 176; development of in Occupied Territories 41; Israeli ban on 49, 124, 176; and peace movement 89–90; and preceding charitable organizations 75; role in *intifada* 49, 176; and social welfare activities 36–37; within PLO 34; *see also* women's organizations
women's conference, first Palestinian women's conference (1929) 108–109
women's conferences: Nairobi Conference (1985) 56–57; West Bank conference on women (1990) 46–47
Women's Health Project 10, 179, 180–182, 195, 196
women's movement: in Occupied Territories 33–58; and political voice 30–31
Women's Organization for Political Prisoners (WOFPP) 124–125, 185–186
women's organizations: new Palestinian 39–40; obstacles to progress 40–41; protest groups 125–126; rivalry between 40; traditional Palestinian 38–39; *see also* women's committees

Women's Studies Centres in Occupied Territories 48
Women's Work Committees, on employment rights 40
Woolf, Y, on sexism and militarism 127
work *see* employment
World Bank, on effect of investment in education 177

Yisraeli, E., on education 111
young people: groups for 55n.; involvement in *intifada* 43–44, 71, 72, 77, 85n.
Youssef, N., on employment of Arab women 138, 139
Yuval-Davis, N. and Anthias, F., on absence of Palestinian state 49
Yuval-Davis, N., on Israeli 'demographic war' 129

Zerubavel, Y., on ideology of sacrifice 130
Zion 2
Zionism 2–3; and colonialism 14n., 35; Jewish opposition to 88; and national security 130–133; and Palestinian nationalism 64–65; use of gender and race 179–180; and violence 132–133
Zureik, E., on discrimination against Palestinian Israelis 113